OPEL GT
1968-1973

Compiled by
R.M. Clarke

ISBN 0 907 073 638

Distributed by
Brooklands Book Distribution Ltd.
'Holmerise', Seven Hills Road,
Cobham, Surrey, England

Printed in Hong Kong

BROOKLANDS BOOKS SERIES

AC Ace & Aceca 1953-1983
AC Cobra 1962-1969
Alfa Romeo Giulia Berlinas 1962-1976
Alfa Romeo Giulia Coupés 1963-1976
Alfa Romeo Spider 1966-1987
Aston Martin Gold Portfolio 1972-1985
Austin Seven 1922-1982
Austin A30 & A35 1951-1962
Austin Healey 100 1952-1959
Austin Healey 3000 1959-1967
Austin Healey 100 & 3000 Collection No. 1
Austin Healey 'Frogeye' Sprite Collection No. 1
Austin Healey Sprite 1958-1971
Avanti 1962-1983
BMW Six Cylinder Coupés 1969-1975
BMW 1600 Collection No. 1
BMW 2002 1968-1976
Bristol Cars Gold Portfolio 1946-1985
Buick Riviera 1963-1978
Cadillac Automobiles 1949-1959
Cadillac Automobiles 1960-1969
Cadillac Eldorado 1967-1978
Cadillac in the Sixties No. 1
Camaro 1966-1970
Chevrolet 1955-1957
Chevrolet Camaro Collection No. 1
Chevelle & SS 1964-1972
Chevy II Nova & SS 1962-1973
Chrysler 300 1955-1970
Citroen Traction Avant 1934-1957
Citroen 2CV 1949-1982
Cobras & Replicas 1962-1983
Cortina 1600E & GT 1967-1970
Corvair 1959-1968
Daimler Dart & V-8 250 1959-1969
Datsun 240z 1970-1973
De Tomaso Collection No. 1
Dodge Charger 1966-1974
Excalibur Collection No. 1
Ferrari Cars 1946-1956
Ferrari Cars 1962-1966
Ferrari Cars 1969-1973
Ferrari Dino 1965-1974
Ferrari Dino 308 1974-1979
Ferrari 308 & Mondial 1980-1984
Ferrari Collection No. 1
Fiat X1/9 1972-1980
Ford Falcon 1960-1970
Ford Mustang 1964-1967
Ford Mustang 1967-1973
Ford RS Escort 1968-1980
High Performance Escorts MkI 1968-1974
High Performance Escorts MkII 1975-1980
Hudson & Railton Cars 1936-1940
Jaguar Cars 1957-1961
Jaguar Cars 1961-1964
Jaguar Cars 1964-1968
Jaguar MK2 1959-1969
Jaguar E-Type 1961-1966
Jaguar E-Type 1966-1971
Jaguar E-Type V12 1971-1975
Jaguar MK2 1959-1969
Jaguar XKE Collection No. 1
Jaguar XJ6 1968-1972
Jaguar XJ6 Series II 1973-1979
Jaguar XJ6 & XJ12 Series III 1979-1985
Jaguar XJ12 1972-1980
Jaguar XJS 1975-1980
Jensen Cars 1946-1967
Jensen Cars 1967-1979
Jensen Interceptor Gold Portfolio 1966-1986
Lamborghini Cars 1964-1970
Lamborghini Cars 1970-1975
Lamborghini Countach Collection No. 1
Lamborghini Countach & Urraco 1974-1980
Lamborghini Countach & Jalpa 1980-1985
Lancia Stratos 1972-1985
Land Rover 1948-1973
Land Rover Series II & IIa 1958-1971
Land Rover Series III 1971-1985
Lotus Cortina 1963-1970
Lotus Elan 1962-1973
Lotus Elan Collection No. 1
Lotus Elan Collection No. 2
Lotus Elite 1957-1964
Lotus Elite & Eclat 1974-1981
Lotus Turbo Esprit 1980-1986
Lotus Europa 1966-1975
Lotus Europa Collection No. 1
Lotus Seven 1957-1980
Lotus Seven Collection No. 1
Maserati 1965-1970
Maserati 1970-1975
Mazda RX-7 Collection No. 1
Mercedes 230/250/280SL 1963-1971
Mercedes 350/450SL & SLC 1971-1980
Mercedes Benz Cars 1949-1954
Mercedes Benz Cars 1954-1957
Mercedes Benz Cars 1957-1961
Mercedes Benz Competition Cars 1950-1957
Metropolitan 1954-1962
MG Cars 1929-1934
MG TC 1945-1949
MG TD 1949-1953
MG TF 1953-1955
MG Cars 1957-1959
MG Cars 1959-1962
MG Midget 1961-1980
MG MGA 1955-1962
MGA Collection No. 1
MGB Roadsters 1962-1980
MGB GT 1965-1980
Mini Cooper 1961-1971
Morgan Cars 1960-1970
Morgan Cars 1969-1979
Morris Minor Collection No. 1
Old's Cutlass & 4-4-2 1964-1972
Oldsmobile Toronado 1966-1978
Opel GT 1968-1973
Pantera 1970-1973
Pantera & Mangusta 1969-1974
Plymouth Barracuda 1964-1974
Pontiac GTO 1964-1970
Pontiac Firebird 1967-1973
Pontiac Tempest & GTO 1961-1965
Porsche Cars 1960-1964
Porsche Cars 1964-1968
Porsche Cars 1968-1972
Porsche Cars 1972-1975
Porsche 356 1952-1965
Porsche 911 Collection No. 1
Porsche 911 Collection No. 2
Porsche 911 1965-1969
Porsche 911 1970-1972
Porsche 911 1973-1977
Porsche 911 Carrera 1973-1977
Porsche 911 SC 1978-1983
Porsche 911 Turbo 1975-1984
Porsche 914 1969-1975
Porsche 914 Collection No. 1
Porsche 924 1975-1981
Porsche 928 Collection No. 1
Porsche 944 1981-1985
Porsche Turbo Collection No. 1
Reliant Scimitar 1964-1986
Rolls Royce Silver Cloud 1955-1965
Rolls Royce Silver Shadow 1965-1980
Range Rover 1970-1981
Rover 3 & 3.5 Litre 1958-1973
Rover P4 1949-1959
Rover P4 1955-1964
Rover 2000 + 2200 1963-1977
Rover 3500 1968-1977
Rover 3500 & Vitesse 1976-1986
Saab Sonett Collection No. 1
Saab Turbo 1976-1983
Singer Sports Cars 1933-1934
Studebaker Hawks & Larks 1956-1963
Sunbeam Alpine & Tiger 1959-1967
Thunderbird 1955-1957
Thunderbird 1958-1963
Triumph 2000-2.5-2500 1963-1977
Triumph Spitfire 1962-1980
Triumph Spitfire Collection No. 1
Triumph Stag 1970-1980
Triumph Stag Collection No. 1
Triumph TR2 & TR3 1952-1960
Triumph TR4.TR5.TR250 1961-1968
Triumph TR6 1969-1976
Triumph TR6 Collection No. 1
Triumph TR7 & TR8 1975-1982
Triumph GT6 1966-1974
Triumph Vitesse & Herald 1959-1971
TVR 1960-1980
Volkswagen Cars 1936-1956
VW Beetle 1956-1977
VW Beetle Collection No. 1
VW Golf GTi 1976-1986
VW Karmann Ghia 1955-1982
VW Scirocco 1974-1981
Volvo 1800 1960-1973
Volvo 120 Series 1956-1970

BROOKLANDS MUSCLE CARS SERIES

American Motors Muscle Cars 1966-1970
Buick Muscle Cars 1965-1970
Camaro Muscle Cars 1966-1972
Capri Muscle Cars 1969-1983
Chevrolet Muscle Cars 1966-1972
Dodge Muscle Cars 1967-1970
Mercury Muscle Cars 1966-1971
Mini Muscle Cars 1961-1979
Mopar Muscle Cars 1964-1967
Mopar Muscle Cars 1968-1971
Mustang Muscle Cars 1967-1971
Shelby Mustang Muscle Cars 1965-1970
Oldsmobile Muscle Cars 1964-1970
Plymouth Muscle Cars 1966-1971
Pontiac Muscle Cars 1966-1972
Muscle Cars Compared 1966-1971
Muscle Cars Compared Book 2 1965-1971

BROOKLANDS ROAD & TRACK SERIES

Road & Track on Alfa Romeo 1949-1963
Road & Track on Alfa Romeo 1964-1970
Road & Track on Alfa Romeo 1971-1976
Road & Track on Alfa Romeo 1977-1984
Road & Track on Aston Martin 1962-1984
Road & Track on Auburn Cord & Duesenberg 1952-1984
Road & Track on Audi 1952-1980
Road & Track on Audi 1980-1986
Road & Track on Austin Healey 1953-1970
Road & Track on BMW Cars 1966-1974
Road & Track on BMW Cars 1975-1978
Road & Track on BMW Cars 1979-1983
Road & Track on Cobra, Shelby & Ford GT40 1962-1983
Road & Track on Corvette 1953-1967
Road & Track on Corvette 1968-1982
Road & Track on Corvette 1982-1986
Road & Track on Datsun Z 1970-1983
Road & Track on Ferrari 1950-1968
Road & Track on Ferrari 1968-1974
Road & Track on Ferrari 1975-1981
Road & Track on Ferrari 1981-1984
Road & Track on Fiat Sports Cars 1968-1987
Road & Track on Jaguar 1950-1960
Road & Track on Jaguar 1961-1968
Road & Track on Jaguar 1968-1974
Road & Track on Jaguar 1974-1982
Road & Track on Lamborghini 1964-1985
Road & Track on Lotus 1972-1981
Road & Track on Maserati 1952-1974
Road & Track on Maserati 1975-1983
Road & Track on Mazda RX7 1978-1986
Road & Track on Mercedes 1952-1962
Road & Track on Mercedes 1963-1970
Road & Track on Mercedes 1971-1979
Road & Track on Mercedes 1980-1987
Road & Track on MG Sports Cars 1949-1961
Road & Track on MG Sports Cars 1962-1980
Road & Track on Mustang 1964-1977
Road & Track on Peugeot 1955-1986
Road & Track on Pontiac 1960-1983
Road & Track on Porsche 1951-1967
Road & Track on Porsche 1968-1971
Road & Track on Porsche 1972-1975
Road & Track on Porsche 1975-1978
Road & Track on Porsche 1979-1982
Road & Track on Porsche 1982-1985
Road & Track on Rolls Royce & Bentley 1950-1965
Road & Track on Rolls Royce & Bentley 1966-1984
Road & Track on Saab 1955-1985
Road & Track on Toyota Sports & GT Cars 1966-1986
Road & Track on Triumph Sports Cars 1953-1967
Road & Track on Triumph Sports Cars 1967-1974
Road & Track on Triumph Sports Cars 1974-1982
Road & Track on Volkswagen 1951-1968
Road & Track on Volkswagen 1968-1978
Road & Track on Volkswagen 1978-1985
Road & Track on Volvo 1957-1974
Road & Track on Volvo 1975-1985

BROOKLANDS CAR AND DRIVER SERIES

Car and Driver on BMW 1955-1977
Car and Driver on BMW 1977-1985
Car and Driver on Cobra, Shelby & Ford GT40 1963-1984
Car and Driver on Datsun Z 1600 & 2000 1966-1984
Car and Driver on Corvette 1956-1967
Car and Driver on Corvette 1968-1977
Car and Driver on Corvette 1978-1982
Car and Driver on Ferrari 1955-1962
Car and Driver on Ferrari 1963-1975
Car and Driver on Ferrari 1976-1983
Car and Driver on Mopar 1956-1967
Car and Driver on Mopar 1968-1975
Car and Driver on Pontiac 1961-1975
Car and Driver on Porsche 1955-1962
Car and Driver on Porsche 1963-1970
Car and Driver on Porsche 1970-1976
Car and Driver on Porsche 1977-1981
Car and Driver on Porsche 1982-1986
Car and Driver on Saab 1956-1985
Car and Driver on Volvo 1955-1986

BROOKLANDS MOTOR & THOROUGHBRED & CLASSIC CAR SERIES

Motor & T & CC on Ferrari 1966-1976
Motor & T & CC on Ferrari 1976-1984
Motor & T & CC on Lotus 1979-1983
Motor & T & CC on Morris Minor 1948-1983

BROOKLANDS PRACTICAL CLASSICS SERIES

Practical Classics on Austin A40 Restoration
Practical Classics on Land Rover Restoration
Practical Classics on Metalworking in Restoration
Practical Classics on Midget/Sprite Restoration
Practical Classics on Mini Cooper Restoration
Practical Classics on MGB Restoration
Practical Classics on Morris Minor Restoration
Practical Classics on Triumph Herald/Vitesse
Practical Classics on Triumph Spitfire Restoration
Practical Classics on VW Beetle Restoration
Practical Classics on 1930S Car Restoration

BROOKLANDS MILITARY VEHICLES SERIES

Allied Military Vehicles Collection No. 1
Allied Military Vehicles Collection No. 2
Dodge Military Vehicles Collection No. 1
Military Jeeps 1941-1945
Off Road Jeeps 1944-1971
V W Kubelwagen 1940-1975

CONTENTS

5	A Rare Opel	Sports Car World		1967
9	GM's Gee-Whizzers	Car Life	Dec.	1967
14	Opel to Launch GT	Autocar	Mar. 14	1968
15	Opel GT Corvette for Europe?	Autocar	Oct. 3	1968
16	GM's (Germany) New GT	Sports Car World	Nov.	1968
18	Opel's New GT	Wheels	Nov.	1968
21	Opel GT	Road & Track	Dec.	1968
24	The Surprising Opel GT	Motor	Mar. 22	1969
26	Opel GT Observed	Autocar	Apr. 3	1969
28	Opel GT Road Test	Motorcade	June	1969
32	1.9 Opel GT Road Test	Road & Track	June	1969
38	Opel GT Big Surprise Road Test	Car Life	June	1969
44	Why the Excitement at Buick-Opel Dealers	Road Test	July	1969
52	Best Import — Opel GT	Car Life	Sept.	1969
54	Opel GT 1.9 Road Test	Car & Driver	Sept.	1969
58	Opel GT 1900 Road Test	Autocar	Sept. 11	1969
62	Opel GT	Road Test	Jan.	1970
66	Opel GT Road Test	Motor Sport	Sept.	1970
67	Opel GT Road Test	Motor	June 20	1970
73	Opel GT Tuning Test	Motor Sport	Dec.	1971
74	Opel GT	Car	Aug.	1970
76	Opels with Hairs on their Chests	Car	May	1971
80	The $3500 GT Comparison Test	Road & Track	July	1971
87	Opel	Road Test	Feb.	1973
88	Engine & Drive — Where Should They Go? Comparison Test	Road & Track	July	1973
96	Tomorrow's Car Today	Asian Auto	July	1976
98	The Opel GT	Thoroughbred & Classic Cars	Nov.	1980
101	Classic Curves	Thoroughbred & Classic Cars	June	1987

ACKNOWLEDGEMENTS

The Opel GT caught my attention a couple of years ago on a visit to the US. It struck me as a pleasing vehicle and I endeavoured to obtain a book that would give me an insight into its background and development. Nothing presented itself, so in desperation turned to Classic Motorbooks comprehensive catalogue, only to find that the only non-technical literature on Opels available was a Car Graphic book in Japanese covering German Fords and Opels. This prompted my search for the articles that will be found on the following pages.

For readers who would like a formal introduction to this car, may I suggest you turn to Jerry Sloniger's excellent article on page 99. He knows the car exceedingly well and was involved with it before its official announcement. His authorative historical piece outlines the highlights of the GT story and brings us up to date. Let us hope that one day he will write a full history of the car.

I was surprised to find that over 100,000 were built in its five years of production. To put this into perspective, approximately 72,000 Jaguar E Types were manufactured during the fifteen years prior to its demise in 1975.

Brooklands Books endeavour to make available lost information for current owners of interesting cars. They have been a service for restorers and historians in one form or another for nearly twenty years. They exist for two reasons, firstly because there is a need, and secondly and more importantly because the publishers of the world's leading automotive journals are themselves motoring enthusiasts and allow their copyright articles to regularly appear in this reference series. I am sure that owners of Opel GTs will wish to join with me in thanking the management of Asian Auto, Autocar, Car, Car & Driver, Car Life, Motor, Motorcade, Motor Sport, Road & Track, Road Test, Sports Car World and Thoroughbred & Classic Cars for their continued support.

My thanks also go to Bruce Joy to Carpinteria, California, who kindly allowed me to photograph his splendidly preserved 1971 GT which can be seen on the front cover.

R.M. Clarke

If you begin a far-out exercise with your engineers and design staff you've got to be very careful. Otherwise you're liable to end up with the world's most swingin' unit ... like this one we'll simply call ...

A RARE OPEL

The German GM establishment built this 140 mph saucer-shape test track. Enthusiastic GM engineers wanted to build a 140 mph chassis to suit the track. Brilliant GM designers wanted to build a 140 mph bodyshell to suit the chassis. Now the car is finished the GM sales force want to sell it. Let's just hope they do it soon.

GENERAL MOTORS is a large firm with a large and world-wide disinclination to indulge in the performance image (odd Holdens and such overseas exercises apart) but there is reason to believe that this iron edict is at least open to negotiation among the branches. The Vauxhall "safety study" shown at last spring's Geneva Salon, sans engine, was one such, as were the Mako Shark and that seven-litre Corvette prototype run at Sebring, if you stop to muse on them.

But the prime goer of the show-car lot is the German GM Opel GT, first seen at the 1965 Frankfurt fall showing, and later all around the motor show circuit in Europe, including Geneva. This one has an engine, and a good one. What's more it might even reach the masses — and I have a hunch that the sales success of cars like the performance-image Mustang will have no little to do with it.

Like that Vauxhall, the original concept of an Opel GT was as a "mobile test bed for suspension work" since Opel was then building what recently opened as the finest rough-going and high-speed test set-up in Germany, if not all of Europe. Despite then-GM reluctance to know about performance, Opel engineers got their GT built on the grounds that "if we have a 140 mph saucer we need a near-as-not 140 mph car, right?" Right.

Next step was the theory that if you have a 140 mph chassis, or something close, you need sleek, streamlined bodywork and the design boys would like some far-out exercise too to freshen their eyes, right? Right. "And then big daddy will let us put it in a salon, right? Probably not, but let's build it, anyway."

As things turned out, that record first-million Mustangs and similar factors like new top brass in Detroit, did allow Opel to show the car. Now

Left: Opel GT wasn't expected to go into production, but the company experiment showed such excellent results, and the cost was so feasible, that it was a natural move.

Above: Like a gull spreading its wings after a flight, the GT shows itself at its aerodynamic worst. High-powered lights reduce top speed by 3 to 5 mph.

Right: GM engineers swooped around the test track at speeds between 85 and 140 mph, and found they could let the car steer itself on the parabolic 45 degree banking.

even Opel can't seem to catch up with a firm rumor that it will be building Opel GTs for you and me with the debut scheduled for Frankfurt 1967 (bi-annual show) and a near-unbelievable price goal of roughly $A2270 (£1135) ex-works. Stern denials at this point, and much clearing of throats at headquarters, but the well-known "reliable source" is so likely to be reliable I'd take a very small bet at even money.

Of course the car as it now stands, with a tuned version of Opel's 1.9 litre Rekord engine, and the price tag may not match. It seems very likely that the cheapest Opel GT will cost that much (my guess), the hot 2 litre somewhat more since these exercises have a way of getting dearer as the months wear on. However, the price leeway Opel enjoys between the target price and the competition in Germany is so considerable they can afford a bit of late-date increase.

By comparison, and using very rough figures translated into your kind of money, the four-cylinder Porsche 912 of 1.6 litres would cost $A4000 (£2000), the 2 litre six with performance to match the Opel quotes a healthy $A5200 (£2600), or twice as much plus a VW for commuting. The 230 SL Mercedes or BMW 2000 Coupes fall between the Porsches, well above the Opel figure, the cheapest GT in Germany with anything like comparable performance is the Glas 1700, 15 mph slower and over $A3000 (£1500) more.

In short, for the money Opel should have a ready-made market for at least the 20,000 it has to sell to turn a profit.

Bearing all this in mind and realising fully that the current design and engineering exercise is far from the last, or even penultimate word on a car we *might* see in 18 months, I still skulked in a corner looking like a potted palm until all the

Deeped dish steering wheel, bucket seats, console and floor shift are all trademarks of a fine, fast car. Instruments are well spread along the dash with small battery of toggle switches in centre.

From the rear the GT just oozes appeal. The entire car was built as a purely functional unit to test high speed handling, but somehow the cart got away with the horse. It would really shake the opposition.

other journalists invited to its test track opening had bused home and then wangled a ride and some photos of the GT. My rationale was: you say the car was built to test the new track which is here to be tested, so let's test, right? Never mind.

Speaking of tracks, Opel opened the first German internal test facility of this kind in 1951 but outgrew it as speeds rose. With no more room inside the factory walls for expansion it found a plot some 20 miles away and built what is again the finest test facility in Germany if not Europe, well guarded by large alsatians and second-growth pine trees, in case a nasty competitor wants to peek.

The operative part is enclosed by a round, scientifically banked track of about 3.1 miles with three lanes of varying degrees of bank, good for roughly 30-50, 50-80 and 80-140 mph respectively, "hands-off" and with no g loads to upset suspension tests. And that was only openers. The rest of the one-way 20 or so miles of road are devoted to such things as six lanes of varied surfaces, including a section of Belgian block suitably worn over the decades and hand-carried to the track to be relaid.

Other efforts laid with equal care include two drag-brake strips, a dead level skid pad, a cobble stone circle which can be artificially dampened, sand and mud runs, and water splashes, fresh or salt to choice, for a start.

Opel even built a man-made hill with gradients of 4 to 30 percent to test clutches on restart and brakes through engineered bends, as well as wavy bits with hand sculptured sine curves which lift the wheels off the ground and the fillings out of your teeth. In short: the ideal of American precision testing inside ten-foot fences.

With some $A68 million involved in these super roads — which you could probably duplicate around any Australian station, billiard-table speed saucer apart — Opel really did need a special handling package as a standard to measure their road cars by. This is the Opel GT.

Since the factory only has one such test vehicle and another show shell Opel was naturally reluctant to turn me loose on the twisty bits (understand they were *polite* with the meat axe when I suggested it) but we did grab some pix on the hill and some impressions around the saucer. Thus handling reports will have to wait — either forever or more hopefully only a couple of years.

The factory has also, so far, eschewed outside

You'd go a long, long way to find a prettier-looking car for the price than Opel's GT. It could be described as stunning, or sleek, or even sexy. Depending on your outlook. Visibility is good.

The normal Opel engine reads 90 (DIN) bhp while the GT version hustles with 125 to 128 bhp on 10 to one compression. But the engine isn't really extended, it'll do 130 mph as it is, but can go more yet.

hatches to the luggage shelf behind the two seats, and for that matter occasional seats themselves, but for the price I don't see many young-at-heart and short-of-purse buyers complaining much.

The two seats that are provided are pure buckets, not fakes, which make safety belts redundant except for safety. You certainly won't need them to stay firm and one with your car.

Most of the current engineering effort has gone into the engine compartment, where there was Opel's latest mid-range Rekord mill to start with. In normal road trim for Herr Jedermann this is a 1.9 (also 1.5 and 1.7 to order) single overhead cam, five main bearing inline four, vastly understressed because it just came out and Opel likes to have latitude for development.

While the normal engine gives 90 (DIN) bhp the GT version at the moment has 125-128 bhp on 10 to one compression with 115 lbs/ft of torque all the way from 3000 to 4500 rpm. In other words, Opel didn't have to get really wild for its rated 130 mph top speed. (A 2-litre Porsche six gives 130 (DIN) bhp by comparison.)

The best indication I can give of its state of tune, is the fact that five or six laps of the saucer flat out can be followed immediately by a halt and it will idle quietly with no signs of baulking. At the moment the car is using four carburettors fed from a common float bowl which would probably prove very thirsty but the retiring state of engine tuning being what it is in Germany any hot engineer could probably get the same performance from a good dual carb with less consumption. Opel achieved its first results — and it stresses this is an early state — by fitting a new head, one number up on compression, better valves and a hotter but not wild cam, along with dual exhausts and an improved header. In short this is a relatively mild tune: about what a Californian would call "good pepper-tree tweak". If anybody wanted to go 2-litre racing there is room for more.

With the benefits of that engine (also available from the parts bin as the Kapitan 2.8 litre single OHC six giving 125 (DIN) bhp dead stock and by extension of the same tuning theory perhaps 170-175 true bhp) Opel could field quite a performance image of its own.

This is all a minor if compared to the major question left today: will Opel or won't Opel? It had better and *if* it does it may well find hotcakes selling like Opel GTs.

#

GM'S GEE-WHIZZERS
EXCITING THINGS FROM GM'S BRAINS ABROAD

BY ERIC NIELSSEN

AMERICAN MOTORS will soon introduce a new 2-seater sports car, the AMX, in the under-$3000 range, and it's said that the Big Three have no counterpart ready. That may be true in the U.S., but not of their affiliates abroad, which can export to the U.S. to the extent required by company policy. Chrysler has its Sunbeams, Alpine and Tiger, and its new 1200-cc Simca coupe, and General Motors has two impressive experimental cars that could take direct aim at the under-$3000 bullseye.

Experimental cars overseas, shown to the public by major manufacturers, aren't the common objects they've become in the U.S. Since GM's Y-Job and Chrysler's Thunderbolt and Newport in the late '30s, "dream cars" have been an accepted part of the American automotive scene. Today, as they've become less flamboyant and more practical—most often fully operational prototypes—they're more appropriately known as "experimental cars."

These not-for-production automobiles provide: 1) Diversionary excitement and cranial exercise for stylists, engineers and product planners, 2) excellent attractions for displays at automobile shows and dealerships, 3) image projections of a company's vitality and ability in car design, 4) information on public reactions to new styling and engineering features of experimental cars.

European automakers (as distinct from body builders such as Pininfarina and Bertone) rarely have built special show cars or displayed non-production vehicles. They feared that such cars would draw attention away from the bread-and-butter models on show, and that the public wouldn't understand or accept the fact that the car was not actually going to be produced. Moreover, European firms traditionally are ▶

DECEMBER 1967

short staffed in the styling department and any extra time or money spent on special cars is usually in racing or record machines—a policy that can hardly be criticized.

Now General Motors, prime exploiter of the experimental car idea in the U.S., has moved to do the same in Europe through its subsidiaries, Adam Opel AG in Germany and Vauxhall Motors Ltd. in Great Britain. In each case the motivations have been different. Opel is a large volume producer, second only to Volkswagen in Germany, and it can see room in its model lineup for a sports car, especially in view of the scarcity of low-priced German-built sports cars.

Opel also felt that a sporty prototype would help focus public opinion on Opel products, which until the last several years have been notable for their unrelieved dullness. Vauxhall faced a similar problem: new and improved products that were bucking a middle-class family car reputation.

Semon E. "Bunky" Knudsen, whose GM responsibilities then included overseas activities, had faced and solved a similar problem at Pontiac. He knew an experimental car like the XVR would help show the public what Vauxhall's engineers and stylists could do.

Though Vauxhall is smaller than Opel, it faces much more domestic sports car competition, so there remained the possibility of a future sports car in the Vauxhall line.

OPEL'S EXPERIMENTAL GT is built on the Kadett floor pan and uses the Kadett independent front suspension. Its engine is the Opel 1897-cc (106 cid) sohc four, modified to produce 108 DIN bhp; up from 90 stock.

GEE-WHIZZERS

Opel's GT coupe was styled under the direction of C. M. MacKichan, one of GM's best designers, who served a stint as chief Chevrolet stylist. This should lessen any amazement over the close resemblance between the GT and the experimental Corvair Monza GT coupe, built in 1962, with its flip-top roof. The similarity is especially strong in the peaked fender lines, cut-off and recessed tail and the use of concealed headlamps for a low flat nose. Work on the Opel GT was carried out entirely at the new styling center at Rüsselsheim by ripening and imaginative young German stylists.

At the front, the GT uses a wide air inlet slot below the slim integral bumper that follows the nose peak line. This requires a fairly high nose—a potential source of front-end lift—which could be reduced by a spoiler just behind the air inlet. The headlamps pop up electrically, and squared lenses reduce the amount of upward travel required for full extension.

Other body features of the Opel GT include doors extending up into the roof for entry clearance on this 46-in.-high car, and access to the rear luggage compartment through the interior. In the rear there is space for minimal seating, with the usual fold-down seatback to give a flat luggage area. The bucket seats and multi-dialed panel, merging into a central console, are very much in the modern European GT idiom. Driving position is very good, and the narrow pillars allow excellent visibility all around.

Opel's engineers used a blend of components in the chassis. The floorpan is basically Kadett, as is the independent front suspension, using a transverse leaf spring. At the rear, they located the coil-sprung axle with trailing arms and a lateral Panhard rod. This layout was new to Opel but since has been adopted on the firm's Rekord and Commodore models.

The basis of the GT's engine is Opel's 106 (1897-cc) four. In stock form, with its camshaft in the head operating in-line valves through tappets and rockers, it produces 90 DIN bhp at 5100 rpm. For the GT, Opel produced an experimental cylinder head with a higher camshaft, lighter valve gear, improved porting, larger valves and higher lift cam profiles. On a 10:1 compression ratio, it produces 128 DIN bhp, giving it more punch than any production sports car in the under-2-liter class (with the exception of the Porsche 911 in its tamest form). The sohc Opel is a high-torque, smooth-idling unit that uses a special Solex carburetion system with four semi-downdraft throats fed by a single float chamber.

Opel has built two running GT coupes, one mainly for auto show display and the other for testing on the company's new high-speed track at Dudenhofen, where the coupe easily clocks laps at 125+ mph speeds. First shown at Frankfurt in 1965, the GT created quite a stir among the German makers, who like to believe they each have cozy market segments to exploit without troublesome and confusing competition.

When the GT coupe was first decanted at the Frankfurt Show's loading dock, a well-known designer spotted it and hastened to the Porsche stand: "There's a new Opel sports car here, and it's pretty nice. You'd better come and have a look." The Porsche reply was unequivocal and final: "Don't be silly; that's impossible. *Opel* can't build a sports car." But Opel had. The little silver car has been shown from Sweden to South Africa and even in New York, to the delight of the Buick dealers selling Opels.

Opel dealers worldwide soon may have GT coupes to sell because Opel supposedly began to tool for limited production of the GT, probably with a fiberglass body, perhaps to be made in France rather than Germany. It would be offered with a wide range of engines and at a price competitive with the British sports cars.

THE OPEL'S interior is a good blend of Detroit and Europe. Instruments are angled toward the driver.

DECEMBER 1967

VAUXHALL'S VXR is the more exotic of the two show cars, and probably the most successful esthetically. Cockpit entry is through large door panel hinged at the unique split windshield centerline.

GEE-WHIZZERS

IF OPEL IS trying to catch up with the British sports car technique, its sister in Britain, Vauxhall, gave that technique a substantial shove forward with the XVR, one of the most radical front-engined sports coupes ever made. It recalls the Bill Thomas Cheetah with its long nose and extreme rearward placement of the engine and occupants. The very slick shape has been called "sleek, sinister," "ugly and menacing," "extremely attractive" and "a brutish beaked monster conceived in a wind tunnel." It lurks close to the ground, only 40 in. high, with the squat proportions that result from an 85-in. wheelbase and 56-in. track.

The XVR is far more experimental technically than the Opel GT. Its chassis is a sheet steel platform, reinforced by a deep central backbone, Miurastyle; by integral seat pans, the seats being fixed, and by a built-in rollbar structure that also supports the door pivots. The rear deck of the fiberglass body lifts to accommodate luggage, while the hood pivots up and ahead to uncover the engine and spare wheel, stowed forward.

Suspension is independent all around, along competition car lines, with parallel wishbones set to give a fairly low roll axis, 3 in. above ground level, minimizing camber change to suit the wide profile racing tires used. These are of 5.00-15 size, mounted on rims 8 in. wide. Centrifugal outlets in the cast aluminum wheels supplement cooling airflow for the outboard disc brakes. These are 9.25 in. in diameter and power boosted. Concentric coil/shock units are used at all four wheels, and the XVR is steered by a rack-and-pinion gear.

Two XVRs were made at Vauxhall's Luton Engineering Centre, one a mockup for show and the other very much for go. It was initially powered

VXR INTERIOR runs more to GM show car styling with exaggerated instrument panel and console.

by Vauxhall's stock VX 4/90 4-cyl. engine, producing 85 bhp from 97.2 cid (1594 cc), with an aluminum head and twin carburetors. The basic design was left very open, however, to accommodate other engines, even a six or V-8. Easily fitted would be the two new fours introduced this fall, a 97.5 cid (1599 cc) developing 82 bhp, and an overbored version of 120 cid (1975 cc) producing 98 bhp. Both engines have a single overhead camshaft driven by a fiberglass-reinforced, cogged rubber belt that also operates a side-mounted accessory package, *à la* Pontiac Six.

Vauxhall's Styling Director, David Jones, used wind tunnel tests with scale models to develop the very clean XVR surfaces, with minimal form drag from the flush-fitting window glass. Interior ventilation is provided by ducts from screened inlets just ahead of the front wheels. Air exits from the cockpit through the recessed rear window, which lowers electrically. Deep under the nose, almost parallel to the ground, are the radiator air inlets. Underhood air is vented out gill-like louvers ahead of the windshield and behind the front wheels.

Among the XVR's other unusual body features are the rectangular headlamps, which are electrically rotated into position around a vertical axis instead of popping up as do the Opel GT's lamps. Radical axes are also used for the XVR's doors, which pivot up and forward along the line of the single central windshield post. This is a scheme stylists have often sketched but have seldom tried full scale, partly because of the supporting strut problem. Vauxhall has neatly solved this with a long arm rearward from the cowl, spring-supporting the door through a ball working in a slide along the base of the door.

Visibility all around is extremely good with this door/window arrangement, though some problems are posed in the windshield wiper department. Vauxhall put none on the XVR mockup, and on the running car they pivoted at first from points close to the center post. After tests, this layout was reversed, with the pivot points outboard instead of inboard.

The XVR carries full circular instrumentation, deeply recessed in the dash panel. All the operating switches and controls, including the shift and hand brake levers, are carried on a panel atop the central chassis backbone. Fore-and-aft adjustment is provided for the suspended pedal assembly, since the seats are fixed, and the small 2-spoke wheel is adjustable for height and distance. Useful hand grips are neatly integrated into the inner door panels.

The XVR was designed at Luton by David Jones and Vauxhall's chief engineer, John Alden, obviously with moral support from GM Styling personnel. After the mockup car was approved, construction of the running XVR was completed in just 12.5 weeks, in time for its debut at the Geneva Show.

Noting that Vauxhall is not now competing in the 2-seater category, John Alden added, "If we do enter the sports car market, we want to get it right." The XVR ably illustrates just how well they could do if they wanted. Will Vauxhall build the XVR or some variation? Alden again: "Obviously behind all these advanced design projects we have some idea that in the long-term future we might put them into production."

A few small photos of the car in an issue of *Road & Track* brought a deluge of requests for more information to GM's Overseas Operations Div. If public opinion were the only criterion, the XVR would be in production already. But Vauxhall is working hard to update its sedan line in Britain's very competitive middle-range market, and it might not have the capacity or funds soon to tool up a sports car. If not, perhaps they'll turn the XVR design over to Lotus. It has the looks the Elan *should* have to match its advanced engineering. Wouldn't *that* be a nice car to own! ■

OPEL TO LAUNCH GT/ABOUT FORD

AUTOCAR 14 March 1968

Three years ago at the Frankfurt show, Opel showed this experimental GT coupé; it is possible that this car forms the basis of the new model to be announced later this year

IT SEEMS certain now that Opel will launch a genuine sports car later this year. As may be remembered, a small coupé with retractable headlights and very rakish lines held pride of place at the Frankfurt Motor Show of 1964. Meanwhile, considerable development work has gone into this project that is chiefly aiming at further improving the "Sporting image" of the Rüsselsheim make that was only a few years ago frequently associated with sloppily sprung Mini-American-type cars with uninspiring performance. Even without the expected GT model, this image has drastically changed with the introduction of new ranges of single OHC engines, revised suspension and modern braking systems.

The new GT, it is rumoured, may become available with a choice of engines between 1,100 and 1,900 c.c. The largest unit in its current form, as supplied with the Rekord Sprint, develops 106 (net) bhp and should provide the GT with a maximum speed of around 130 mph.

More recent results of rallies and speed events seem to show that this General Motors subsidiary starts to take at least a semi-official interest in serious competition. With the eventual manufacture of the GT this trend may become even more pronounced. The Opel GT is plainly aimed at a market that has never before been tapped by the big manufacturers. There is every indication that the GT if not just a cheap car, will at least not be a forbiddingly expensive one. Guesses are that at a price around 10-12,000 DM (£1,040-£1,240) there would be quite a demand for a "standard version" with perhaps a "hot" one at something like an additional 1,500 DM (£156).

About Ford

THE advent of the Escort, which eventually is also to be manufactured at the Belgian Genk plant—hitherto main source of the 12M/15M Cardinal descendants—has led to some speculations about the future of the front-wheel-driven Cologne models. Authoritative sources at the Cologne factory were quick to assure that there was no intention to replace the 12M/15M series cars by the rumoured "teutonized" Escort, which then might be called anything from 10M to Eifel, with the latter in no connection to the famous Paris spaceframe structure but rather the Eifelhills of Nüburgring fame.

The nomenclature of the German Ford models has over the years become quite confusing. Originally, the designations used to give a clue to the capacity of the engines: 12M meant 1.2 litres and 20M used to have a two-litre power unit. The "M" in each case stood for what the Cologne manufacturers thought about their own product: "Masterpiece". More recently this system has started to deteriorate. Although the models seem still to be regarded as masterpieces (1967 production figures seem to raise some doubts here), the 12M is no longer a twelve hundred, but sports a 1,300 c.c. engine; the 20M may be had with a 2.3-litre unit, and for certain markets the prestige version of the smallest model range, the 15M, may even be specified with the 1,200 c.c. unit that "officially" has vanished from the programme. This latter option seems particularly baffling because 12M and 15M cars were fitted with differing grilles, headlights and rear lamp clusters specifically to provide all the Müllers and Schulzes with the required visual information about who could afford what. . . .

Opel GT Corvette for Europe?

NEW FOR '69

Fastest-looking Opel yet, the Opel GT Coupé shape (in original form) first appeared at the 1965 Paris and Frankfurt Shows as a styling exercise. The headlamps retract when not in use

WHOEVER said that General Motors did not like motor sport? It can't be Zora Arkus-Duntov, the man behind GM's unsuccessful attempt to win Le Mans with their Corvettes last week-end. It can't be Pete Estes, Chevrolet's captain, who now promotes nothing else but the "Chevrolet sports department". Could it be Ed Cole, GM's new president, who hired Olympic champion Jean-Claude Killy at $100,000 per year, in order to identify the corporation with sports and associate with an image of victorious agressiveness?

It certainly can't be Ralph "Red" Mason, Opel's president. Officially, of course, Opel would not admit to any participation in racing. Any Opel rally entry—European champion Nasenius in Finland, Lambert in Germany, "Monte" winner Greder in France among others—is still very carefully presented as a "private entry", while every insider knows that such fully fledged pros would drive nothing but a factory-prepared car, and take part only after signing fat *bona fide* contracts with wealthy and well organized racing stables.

Racing results are being used to full advantage by Opel's publicity and promotion men, all over Europe. Following in Fiat's footsteps, Opel were fast to bring sports into the show-room also: the "Kadett Rallye", ever since it was introduced in the US has been Opel's American best-seller. Opel hold second rank—beaten only by VW—as American's most successful importer. The Commodore has done more for Opel's image than any other car ever produced at Russelsheim. Then came such European style "specialty cars" as the Rekord "Sprint" and the Commodore "GS".

"We have still better things on hand," Ralph Mason said a year ago. "The type of car which will make us popular even with the most sophisticated guys in your writing bunch. . . ."

This "better thing" is here, and the first man to drive it publicly was none other than Jean-Claude Killy, when he ran it around the *Circuit de la Sarthe* at Le Mans, shortly before the start of this year's 24-Hours race. To insiders, of course, the Opel GT Coupé is nothing really new: it was first seen in its original form in 1965, at the Frankfurt and Paris shows. At the time, however, the car was nothing more than a showpiece, a styling exercise of the type indulged in frequently by GM designers. The car was not meant to be produced. Nor was it to be driven. It was made for the shows, as one of a long list of dream-cars which never come true.

This one did, however, as Europe and America will soon find out. Inspired by Fiat's success in the field of sporting production cars, and convinced of the promotional value for the whole line of an attractive (and cheap) sports car, Opel spent almost three years redesigning and trying to make their show-car operational. It will be launched at the Paris show as the first "Common Market" sports car. While using mechanical parts imported from Germany, the GT coupé is produced in the suburbs of Paris, by France's largest *carrossier*, Brissonneau & Lotz. The firm already assemble Matra's 530 and various small production Renault models.

To prepare for production of Opel's first sports car, Brissonneau had to build a brand new addition to their plant. Optimistic plans call for a yearly minimum output of 10,000 to 15,000 units, half of which are to be sold eventually in the USA—where Buick prepare to launch the car as GM's answer to Porsche!

The coupé will be available in two mechanical versions. The cheaper of the two (not to exceed $2,800 and therefore underselling Fiat's 124 Coupé within the Common Market) uses a 1,078 cc. engine, derived from the Kadett Rallye 1100 power plant. Power of the engine is 60 hp net or 67 hp gross at 6,000 rpm. This version will achieve a top speed of approximately 100 mph.

The more powerful version is driven by Opel's four-cylinder 1,897 c.c. engine, with overhead camshaft, which produces 90 bhp net or 102 bhp gross at 5,400 rpm. Performance is close to 125 mph. Planned sales price in Europe is less than $3,600 (some £1,500).

Both cars have independent front suspension, servo-assisted brakes (with a double braking circuit and forward disc brakes). Presentation of these very low-profiled cars is very sporting, with two racing seats (a third passenger could possibly sit in the back), a completely black instrument panel, a very short gear-shift lever and retractable head-lights.

Russelsheim thus provide a European answer to Detroit's Corvette. Chevrolet, of course, won't mind. But Fiat, Ford of Germany and a few others in Europe are bound to be startled.

Edouard Seidler

Subsidiary instruments are angled towards the driver; the entire panel is black-finished. At the moment, only left-hand-drive versions are available

Businesslike backside. The car is available with 1.1- or 1.9-litre engines

Spare wheel is stowed away from inside the car

GM's [GERMANY] NEW GT

If Opel can come up with swinging machinery like this GT — and they've produced an Opel Rekord similar to our Monaro — what chance that GM Australia has something? Here's what Sloniger thinks.

FROM fold-away "frog" lights to twin tail pipes with chrome tips the Opel GT comes on like a swinger, a teeny-bopper Corvette with peekaboo slits all over its facade.

But there comes a time when you have to quit the ogle and pull it over your thighs like a pair of tailored jeans. You're ready to play rough. This is the moment you discover that Opel's latest offspring speaks with an autobahn accent. Like a society girl on the town, it promises more than it can deliver — at least in 1100 form.

When I drove the car it was chassis number one off the pre-production pilot line. The only kind they had came with 60 DIN hp and 1.1 litres; good for 95 mph in top but slower off the drag-strip line by virtue of an extra 200 lb.

Next on our list is the 90 hp 1.9 litre kind which should do 115 and sell for less than $4000 in deluxe trim. The 1.1 with basic trim — this car — comes in for $2500 and some change at home. Eventually options will include those two engines and plain or fancy everything else.

We had to get along with enormous tachometer and speedometer plus one more dial for water temperature and fuel. Period. The deluxe carries those as well as oil, ampmeter and a clock. Secondary dials are angled toward the driver from the console centre, above a row of tumbler switches. The main faces, about steering-wheel size, are right up front and centre.

The wheel itself — woody sort, naturally — is smaller than most cars these days, requires only three turns lock to lock and operates very light steering. A stubby gear lever a hand's breadth away atop the tunnel works four quickly-found cogs. Then you have a special lever, like a panic handle, to flip main headlamps into view.

Seating is beyond belief for a production car built to fill the vacant middle-price sporting gap around Germany. They are flat and low but rolled deeply for both thigh and shoulder support. The backs (adjustable) reach way up to cushion whiplash but also block vision to some extent.

Looking forward, the driver sees knife-edge wing tops for easy parking while rear quarter view is bad. The wheel arches are flared widely and the cutouts obviously too big for 155 x 13 rubber, so larger tyres would be no problem. Except for the spare. If ever a car needed the Firebird-style deflated spare, this is it. The current ones use ¾ of that thin bin under a fixed rear deck. Clamber past the high seat backs and lift a vinyl flap. Store your socks and put the other luggage on a shelf behind the seats: about two medium cases worth. They wisely made no concessions to 2+2-dom. It is a pure two-seater.

One without interior storage too. A slit before the passenger will take two gloves or the owner's manual. There are no door pockets, tunnel bins or the like.

With doors cut into the top the entry/exit game is easier than most in the class and there is foot room enough, if not lavish space for big brogues with the engine pushed back.

Despite that concession to weight distribution and bonnet height the car understeers badly in 1100 form, without the torque to break your Cinturatos loose even in the wet. The 1.9 engine would furnish throttle steering but it's heavier too. Apart from this tendency to exit curves nose first when pushed, the GT holds a pre-selected

---

SPECIFICATIONS

Engine: Inline four, 1078 cc (75 x 61), 9.2 compression, 60 DIN hp/5200 rpm (68 SAE/6000), twin Solex downdraft carbs, side cam, three main bearings, ohv, alternator.

Chassis: All-independent suspension — wishbones and 3-leaf transverse spring in front, beam rear axle on coils and two trailing arms with Panhard rod. Disc/drum brakes, rack and pinion steering.

Pertinent Facts: Wheelbase 95.7 in., front/rear track 49.4/50.6 in., length 162.0 in., width 62.2 in., height 48.6 in. high, kerb weight 1863 lb. Turning circle 35 ft 4 in., three turns lock to lock. 11 gal tank. Consumption circa 25-28 mpg with 3.89 final drive. Top speed 95 mph. Acceleration: 0-30, 4.9 sec.; 0-40, 8.0 sec.; 0-50, 11.1 sec.; 0-60, 16.3 sec.

---

line well, darts nicely and avoids back axle bounce which afflicted original Kadetts. With quick, accurate steering and solid braking it is entirely safe and fun to flog around the bends. And amazingly elastic for its 1078 cc.

The styling, originated by one American at German GM and put into production by another, is the car's prime feature, even though it lost some pure prettiness between prototype and production. Wind tunnels and marginal luggage needs as well as space for upright carbs dictated higher nose and tail. Production models look more purposeful and have far better cockpit layouts than the original toy.

For only two-thirds the money of a Porsche, and with a lot better quality than any imported two-seater, you can wow the sport car crowd. The name of the game is Opel GT.   #

The car in our lead photo is the prototype of 1966. Compared with the production model we see rounded headlight covers, hood bubble, different grille treatment. Basic concept is similar.

Opel GT interior is very business-like. Two round dials dominate the facia, while all other instruments and gauges are on centre console. If Opel can make it, why not Australia?

TRUISM - OF - THE - MONTH - CLUB: Successful sport cars are erected; seldom born full-blown. The operative word is "successful". Any man might doodle a beautiful shape and panel benders among us could probably mould it around a twin-carb powerplant. But such stardust does not a sport car make.

For that laudable end you must successfully combine a cannily re-rigged production chassis with proper stop/go potential; a careful engine choice from some wide-ranging parts bin and, above all, the sleekest, meanest possible two seater shell a couple of top-notch designers can devise.

Call it an Opel GT. Magazines and the truly shriven fans will drool or knock your toy for its understeer, 0-60 time and/or steering accuracy. But the honest among them must admit that success or failure of such a new, hopefully large-production, sports car rests at last on the sensuous flare of a front wing or perhaps a tail piece to remind all men over 16 of the latest Swedish movie.

Call it an Opel GT. The production GT from German GM is not as pristine and pure as its prototype, but then they seldom are. Wind tunnel efforts, the need for a minimum of spare shirt space in back, and room for production downdraught carbs under the bon-

An Opel that's not quite a gem...

OPEL'S NEW

Sloniger — who tested it two months before release date — reports that if you have a yen for storing things then the Opel GT is definitely not your car.

net dictated higher nose and tail. Regular road runs broke up the slippery front panels with air intakes, outlets and bulges.

The result is a 1.1-litre coupe capable of 95 mph (with a 1.9-litre to follow that will top 115 mph) that is heavier than the comparable Rallye Kadett and thus somewhat slower off the line. This is typically German.

Continued

GT

Continued

They want autobahn beasts — not drag racing. When I made my run there was no choice of engine. In fact there was no choice of cars: we were two months ahead of the release day, driving Pilot Line Model No 1. Options by the time you read this will include a 1.9 ohc engine and deluxe interior for either size car: four possible variations in all.

What all this proves is simply that the 1.1-litre Opel GT looks a hell of a lot meaner than it is. Like the teaser for next week's blue film epic, the vented, louvred, power-bulged GT titillates without the erotics to back up its promise. Maybe that is the way Opel planned it. On its own merit alone the GT is no slouch, mind you. No 100 cc car turning so near the ton, restarting, two-up, on a 30 percent wet cement grade without fuss or covering 300 effortless miles on a tank can be called a slouch. Specially when it looks so right.

Let's take form first: The latest Opel certainly has some family ties with the Corvette — and a sports car could do far worse. Flared, knife-edge wings up front, the long bonnet, Kamm tail and doors cut into the turret all evoke GM day dreams. Those doors even make it possible for anybody young enough to want a GT in the first place to enter or exit in dignity. And the peaked front wings let the same driver see where he is parking. Solid rear ¾ panels and high seat backs do less for rear vision, but then that's the two-seater price.

In point of fact, Opel nudged the ante up a little higher than usual when it came to paying for rear vision by fitting seat backs designed to reduce whiplash but blind as night over your shoulder. I hate to knock those seats — even for blocked vision — because they offer just about the finest production sport car fit in motoring, and that's worth just about anything short of a 5-ton truck up the chuff. (Particularly since an 11-gallon petrol tank comes dead last in the Opel GT).

You enter this GT feet first and settle into low, flat leatherette seats with deep thigh and shoulder rolls that work. They don't seem to hinder entry but sure benefit fast drives. In fact the whole cockpit, set up entirely for two people and damn the friends, looks tighter than it is. Despite an intrusive engine and high tunnel there is ample foot room. The steering wheel is smaller than usual (naturally woody) and a shade high, but it still doesn't block two enormous white-on-black dials for tach and speedometer. The latter has a tenths odometer — but no resettable trip. Opel is prone to these small slips.

In the plain-jane model there are three faces atop the centre console angled towards the driver. Only the centre one features needles — for water temperature and fuel. The left circle holds oil and amps; the right an Opel emblem. In deluxe trim these are replaced by needles and a clock respectively. Below that trio you find a row of tumbler switches and then a high tunnel-top with no tray but a stubby shift lever just clear of the wheel and easily reached. Thankfully the handbrake rests behind the shift — avoiding Opel's penchant for umbrella grips.

A unique feature of the GT is a handle looking like a panic grip for your ejection seat and located near the dash. Moved firmly forward it rolls the main headlamps into position, nothing more. If you yearn to store things, the Opel GT is not your car. There are no door pockets, no tunnel bin and near as dammit no glove locker. A slit in front of the passenger will just about take the operator's manual.

Behind the high seats a strong man can fit in perhaps two medium size cases open to all passing eyes. Still further back there is a thin recess covered by a vinyl flap. Apart from the spare wheel — you wonder if any Opel engineer had to fit a muddy flat in place of his clean spare — there might be room for your short-handled tooth brush. Nothing more. This boot, if it might be dignified by such a name, is only available through the cockpit. At the other end the engine is readily viewable but its hindmost cylinders are hidden under the scuttle. Most aggregates remain available, including a battery ahead of the radiator. Bonnet height dictated a tortuous air cleaner. Sidedraught carbs cost too much.

Quite obviously this engine arrangement is aimed in part at weight distribution. Even so, and using only the smaller 1.1-litre powerplant, the car is a pronounced understeerer. What's more it doesn't have the power to break its Cinturatos loose — even in the wet. So all results of going too fast will be nose first. On the other hand, the 1.9 engine option may provide throttle steering — but it will weigh more up front as well.

Apart from this inclination to charge the scenery like a trained bull if you overdo the loud pedal, the GT is a pleasure to drive. Steering is remarkably light, more than usually precise and Opel has tamed the dreaded back axle tramp so that even Belgian blocks don't catch the car off balance. Brakes are more than up to 1865 lb of kerb weight. Through the gears the Opel GT does 30, 45 and 65 — using the yellow line of 5800 rpm. The big tach reads red from 6100 on, but even then there is no sign of frantic valve dodges. For hot (sic) drag starts you do best with 3500 off the line and then a heavy right foot. Taking the engine to 6000 and dumping the clutch merely causes it to bog down and sigh deeply.

Most important, when you are seated in a shell so sleek, there is a proper throaty exhaust note on acceleration and a quick, not quite nervous, steering response to give the full and proper boy-racer syndrome. What more can a man want at roughly $2600 for the basic-fittings 1100, ex-works? This is about a third more than the 1.1 Rallye Kadett of similar power. But the one-upmanship is worth it. The deluxe 1.9 GT should cost around $3500. That's still 20 percent less than your cheapest Porsche at home, with equal performance.

In fact, Opel and its GM brethren around the world, are aiming directly for the youth market Porsche and friends priced themselves out of. They have the pace and more than enough grace.

Call it an Opel GT. #

OPEL GT

Production version of a show car loses something in the translation

Show car (above) had a longer and smoother nose, was lower at roofline and seems less dumpy and lumpy than production Opel GT.

THREE YEARS AGO Opel surprised the public at the Frankfurt Motor Show when they introduced an experimental car with a definite sports car look. The car was shown to test the public reaction to the car and the reaction was undeniably favorable. Now, three years later, the production version of this experimental design has been introduced—the Opel GT.

Based on the Opel Rallye Kadett, the new car will be marketed in two forms, one with the 1.1-liter engine, the other with the 1.9. The car has the same 95.1-in. wheelbase as the Kadett and shares all the other basic dimensions except that it is slightly shorter (162.0 vs. 164.6) and considerably lower (48.2 vs. 53.9) than the Kadett.

The body of the new car, which is manufactured by the French company of Brissoneau and Lotz, was developed with the assistance of wind tunnel experiments to give it an efficient aerodynamic shape. Because of this, the headlamps disappear into the fenders and are only moved upwards when the lights are to be turned on. This is done by a lever mounted on the gearbox hump. Below the wrap-around bumper there is a grill with two driving lights.

The doors are cut into the roof slightly to make it easier to get in and out. There are bucket seats with high backs and

DECEMBER 1968

OPEL GT

Headlights of show car version (above) lifted from front while production version lights rotate sideways 180 deg.

these are adjustable for both reach and rake. The luggage space is behind the seats and the spare tire is also in the rear.

The instrument panel is deeply padded with the instruments themselves set into the panel. There is a tachometer in addition to the speedometer and other normal dials.

The engines in the two versions are the 1.1 ohv and 1.9 ohc 4-cyl designs used in the Kadett Rallyes. The 1078-cc engine has a 3-main-bearing crankshaft and is rated at 67 bhp while the 1897-cc engine has five main bearings and an output of 102 bhp. A 4-speed manual gearbox is standard in both versions but a 3-speed automatic will be available for the 1900.

The suspension of the GT, also carried directly over from the Kadett, consists of upper and lower A-arms with a transverse leaf spring at the front and a live axle on coil springs at the rear.

Performance figures given by the factory show a top speed of about 95 mph for the 1.1-liter car and 115 for the 1.9.

Only the 1.9-liter version is expected to be exported to the U.S. and it should make its appearance there in early spring. No prices have been announced for the U.S. but a rough guess is that the Opel GT 1900 will sell in the $3500-4000 price range in the U.S.
—*Gunther Molter*

GS 400, Stage I

Opel GT

Buick introduces automobiles to light your fire.

Buick introduces the Opel GT. Who else but Buick would?

Buick introduced you to the Riviera GS, California GS and GS 350.

Buick brought you the GS 400 and Stage I.

Buick introduced you to the Opel Rallye Kadett.

Now Buick introduces the Opel GT, and you have every reason to be excited. Only, please. Have patience. You may not be able to get one right away.

Like all good things, the Opel GT is available in a limited quantity.

Which, if you'll think about it, is really the way it should be.

Automobiles like this don't come along very often. Especially automobiles equipped like this that nearly everyone can afford.

There really isn't much you could ask for that isn't already standard on the Opel GT.

A 67 horsepower engine is standard on the Opel GT. A 102 horsepower engine is available for the most enthusiastic enthusiast.

Power-assisted front disc brakes, dual exhausts, and radial ply tires (165 HR x 13) are standard.

A short-throw, 4-speed stick is standard. Mounted in a console on the floor. A fully automatic transmission is available. Mounted on the floor.

Specially-contoured vinyl bucket seats are standard.

A tach, amp and oil gauges, and 0.1 mile odometer are standard.

And, of course, a very lengthy list of GM safety equipment is standard, too.

There you have it.

The beginning of an exciting story that can be heard in full at any Buick-Opel showroom. Brought to you by you know who.

Who else?

BUICK MOTOR DIVISION

There's a long warm spell ahead.
The light-your-fire Buicks: Riviera GS. GS 400. Stage I. GS 350. California GS. Opel Rallye Kadett. Opel GT.

continental diary

by Paul Frere

Very much like the original Opel experimental car which one of 'Motor's' staff tried at Dudenhoven several years ago.

The surprising Opel GT

FOR some years now the management of Opel has belonged to the clan of GM executives who push as hard as they dare to establish a sporting image for the cars they produce. Perhaps they need it more than most, for although many years have passed since the days when the roadworthiness of Opels was about the worst that you could buy, the memory is still floating about in the public mind. The only radical way to kill the legend is to prove by every possible means that an Opel can also be a well mannered and safe driver's car. A competition programme which has already earned the firm an outright win in the Tour d'Europe, first place in the Touring car class in last year's Monte Carlo Rally (second this year) and a magnificent performance in the Spa 24 Hours is certainly the best way to drive the point home and the sporting range of Rally Kadetts, Rekord Sprint and the 115 m.p.h. Commodore GS is beginning to change the image.

The latest effort in this direction is the 1900 c.c. GT Coupé which has finally matured from a prototype exhibited at the Frankfurt Show in 1965 and is now in production. Following its introduction at the Paris Show of last year, and even before the car could be tried, the demand has been such that the original production facilities have been doubled to 120 a day. The body shells are made by Chausson in Paris and the trimming and painting, which originally was to be done exclusively by Brissoneau and Lotz, also in Paris, will now be executed by Opel's Bochum factory for about half the cars produced, to increase production.

The shape of the GT has lost some of the finesse of the prototype in favour of more realistic inside dimensions and the tail certainly looks a bit clumsy. The general lines and details are, however, the outcome of extensive wind tunnel tests at Stuttgart University and they are certainly very efficient both for reducing the drag and for minimizing wind noise. This, in fact, is virtually non-existent, up to the GT's maximum speed which should be in excess of 112 m.p.h. (I could not time the car accurately, but this is a fair estimation from corrected speedometer readings), and this again reflects the wind cheating ability of the body, for it is over 10 m.p.h. higher than the (lighter) Rally Kadett and 16 m.p.h. higher than for a Rekord saloon with exactly the same engine.

One is tempted to ask why, in fact, the standard 1900 engine (mind you, it *is* an o.h.c. unit) is used in this sporting coupé when a hotter, twin carburetter version is fitted to the Rekord Sprint. The answer is mainly that the second carburetter would have necessitated an enormous power bulge on the bonnet. And frankly, as a first venture, the car is quite fast enough as it is. Perhaps some day we shall see a competition version in which an untidy bonnet would be less objectionable or in which fuel injection would solve the problem, but then, a limited-slip differential would be needed.

This was found going down to Sospel in the course of an entire day's driving in the South of France. Out of the hairpins, acceleration was limited by wheelspin, in spite of the fact that the rather heavy engine unit (it is all cast iron, as well as the gearbox) is located well back in the car, ensuring that about 50% of the weight is borne by the rear wheels with two up, as I drove it.

Otherwise, the GT (which is also available with 1,100 c.c. Rally Kadett engine) is extremely well behaved and should go a long way to change public opinion about the roadworthiness of Opels. Except for spring rates, damper settings and wider (5-inch) rims, the GT's running gear is identical with the Kadett's having a rigid rear axle with a short torque tube and radius rods, located laterally by a Panhard rod. Coil springs are used and, at the front, the lower pair of wishbones is connected to a two-point mounted transverse leaf spring, giving an anti-roll effect. With radial tyres, good weight distribution and rather stiff springs, both handling and road holding are excellent. On faster bends, the car is almost completely neutral and cornering speeds quite high, while in the slower variety, there is usually enough torque available to hang the tail out at will if the surface is at all slippery. But understeer of any severity is never noticeable. Coupled with the light and very positive rack-and-pinion steering and a low polar moment of inertia, this all combines to make up a car which is extremely pleasant and enjoyable to drive.

Rough road surfaces do not disturb it unduly but the standard of comfort provided is definitely in the sporting category and potholes or dips will produce jolts which the seats (the latter could do with more lateral support) will only partially absorb. Perhaps it was a mistake to compromise with the relatively narrow track of the Kadett as the same stability could probably have been achieved with better comfort if a 2- or 3-inch wider track had been adopted. At speed, the car remains very stable, however, and takes 100 m.p.h. bends, creating noticeable lateral g, in its stride, just going where it is pointed without any fuss.

The gearbox, too, works very well though it could do with higher intermediates (specially first and second) than the standard Opels and, at high revs, there is an annoying vibration from the reverse catch. the least refined part of the car is perhaps the engine which can be felt, and heard, to be a rather big four-cylinder. Its main virtue is its high torque in the medium speed range, which does allow the car to be driven in a leisurely way without much sacrifice in performance, but again this characteristic makes one wish for higher 1st and 2nd gears, the latter allowing 53 m.p.h. at the recommended 6,000 r.p.m. limit. At the top end, power is felt to drop beyond the 5800 r.p.m. mark and there would be definitely no point in using more than 6000 r.p.m. in the gears. This is also the figure reached at the car's maximum speed, showing it to be quite notably undergeared, the maximum net power output of 90 h.p. being produced at 5,100 r.p.m. Although it's not impeccably smooth, the engine never feels fussy. In fact, the stopwatch proved the car to perform considerably better than it actually feels: 60 m.p.h. is reached in less than 10 seconds and only 31.9 sec. are required to cover the kilometre from a standing start with spinning wheels, the speed being then very nearly 100 m.p.h., which easily beats a Porsche 912 or a 1967/68 Triumph GT 6. From very high speeds, the servo operated, but very progressive and responsive brakes are a bit marginal however, though curiously, no fade was experienced driving fast down mountain roads.

The body is strictly for two with room for only scanty luggage behind the seats (which have reclining squabs) and for a few oddments in the stubby tail next to the spare wheel. The latter can be reached only from the inside, a messy arrangement for a car that will appeal not only to die-hard enthusiasts but also to women. There are no front quarter windows (nor are they needed) and the rear quarter windows are fixed, but adequate ventilation is provided at all levels. The heater is excellent and an electrically heated rear window is standard.

Window winders which are both hard to operate and difficult to reach figure on the debit side and the lighting, comprising retractable headlights operated by a hand lever, is only fair for a car of that performance, in spite of a pair of fixed small quartz-iodine lamps which also permit flashing with the main lamps retracted. On the other hand, well laid out and elaborate instrumentation (but without a trip recorder), cleverly disposed pedals and the well placed, small steering wheel appeal to a keen driver. A rest for the left foot would be a welcome addition, however.

On the Continent, the fast Commodore GS has already done a great deal to raise the Opel image to a higher level. The GT should move it up another step, for it certainly does not pass unnoticed and provides real fun for a modest outlay. A well driven example is certainly a hard nut to crack where handling is as important as sheer performance.

Smooth, sleek and with the wheels well out—looks fun.

(Continued from page 57)

There is room for several bags back there and they are surprisingly accessible. Not quite so strategically placed is the spare tire which is hidden behind the luggage area above the rear axle and separated from the luggage compartment by a snap-down vinyl flap. If you have to use the spare pray that it's not raining—dragging a muddy tire through the passenger compartment prompts evil thoughts.

Aside from the space behind the seat there is very little room allotted for the necessities of travel in the GT's interior. One open bin in the instrument panel and that's it. No locking glove boxes and no door pockets.

And then there is one other handle, the one sticking up out of the console like a junior-sized shifter which is charged with the responsibility of hiding the headlights. It's a manual operation, of course. What do you expect for $3500? With a mighty push half the front end twirls around and the Opel has eyes.

Unfortunately the headlights don't always latch into position and you have to either go through the whole operation until they do or get out and nudge them with your finger. Porsche would never do it that way, but then maybe the blame lies elsewhere. Even though all Opels have their birth certificates stamped in Germany, 90% of the export models are assembled in Antwerp, Belgium and the GT's coachwork is actually made by Brissonneau et Lotz in France. Considering the overall high quality of the body work and interior trim we would say that the French are holding up their end of the bargain better than the Germans.

Upon seeing the Opel GT for the first time we were prepared to welcome it into our garage with the same enthusiasm that we reserve for only the few really exciting cars that happen along each year. After driving it we have to admit it's a whole lot more Opel Kadett-like than we hoped it would be. If you can overlook the clattering, uncooperative engine, the bad accelerator location and the uninspired handling, which is asking a lot, the GT is a very competent design. It's roomy and comfortable inside. The ride is on the harsh side and the interior noise level is fairly high at cruising speed, but that is true of most of the other similarly priced sports cars. The rack and pinion steering is light and lively and you already know what we think of the transmission. The potential is there. The styling package alone puts it a light year ahead of most of its competitors. With an engine transplant and some serious chassis tuning there is no reason that the Opel GT couldn't be every bit as desirable in the medium-priced sports car world as the Fiat 124. Right now it's just another pretty face.

A closer look at OPEL GT

AUTOCAR, 3 April 1969

THE Opel GT shape first appeared at the 1965 Frankfurt Show as one of those "Purely a research" (pronounce as rhyming with tree-church) "vehicle, man, like to keep the styling boys from chewing their modelling wax, kinda three-dimensional market probe module, man, like the ad man told me. Make it? We wouldn't dream of making it. How *would* you make it"—sort of cars. All American manufacturers delight in playing this game. It helps to distract too close an inspection of a dull standful of frantic-head-scratchingly revamped models which you are furiously telling everybody is New New New when everyone can see that they're at least five-years-Old Old Old and out of date then. General Motors are great dream car players but have lately been cheating by first of all having some striking and sometimes beautiful dreams and secondly having the effrontery to actually make them into production cars. That marvellously uninhibited Mako Shark became the latest Corvette—the Open GT is very like *die Franfurtertraumwagen* of four years ago—the main difference outside is that instead of flicking up rectangular-lidded headlamps as then, it now rolls them over at you, like some amorous West Indian cow-frog coming over coy. Like so many baby frogs, the clear unspotted skin of the child didn't quite survive the four years of adolescence and the necessities of marriage to a production line—a noticeable wart has gathered over the carburettor, above the air intake slits by the bridge of its nose; and the low shark-like mouth of the original has become an interesting case of lifting upper lip caused by ingrowing spotlamps. Nevertheless, the adult is undeniably handsome and shapely.

General Motors Ltd—GM's London "branch"—kindly let us have a closer look at a 1.9-litre Opel GT recently. Though the car was brand new and could not therefore be "figured", let alone driven in the sort of way a sports car should be driven, the experience was interesting. Opel have only recently been involved in any form of competition—absolutely unofficially of course —various models having done quite creditably in European rallies. One doesn't associate sporting characteristics with the marque and it therefore came as something of a surprise to find that the new GT seems to be rather more than a tarted-up body hiding very ordinary internals. Such cars usually give themselves away by over-styled details—the chromium grille over the air-intake that isn't for instance—but there is nothing of the sort here.

The car *is* based on the Opel Kadett. It has the same basic chassis and suspension—traverse-leaf-spring and wishbone i.f.s., rack-and-pinion steering, coil-spring and radius-arm live-axle rear end— and the same engine options (1.1 and 1.9-litre four cylinder ohv and ohc units respectively). The 1.1-litre gives a claimed 67 bhp (net) at 5,400 rpm and the 1.9-litre 102 bhp at the same revs. The gearbox is four-speed all-synchromesh, final drive ratio being 3.89 with the little engine and 3.44 with the big one. Tyres are 165H-13in. and there are other options such as automatic transmission, heated rear window and alternator, and a limited slip differential. The disc-drum brakes have dual hydraulic circuits and servo-assistance; through-flow ventilation is provided, with the extractor vents over the back window. At just below 2,100lb kerb weight, the GT

It's a pretty little animal with a quite dashing blend of sweeping curves, sharpened lines and bulges over an overall length which is just long enough to get away with it. The affinity with transatlantic relatives is clear. The front bumper is really only a capping over the nose but over-riders are fitted; the half-bumpers at the back are more functional. Hazard warning flashers are standard. The spoiler lip over the stubby tail is not big enough to be aerodynamically effective. The Opel is no sluggard in 1.9-litre form—115 mph top speed is claimed

OBSERVED

by Michael Scarlett

Opel's sports car

is around 200lb heavier than the Kadett Coupé, 7in. lower and 2½in. shorter.

The interior generally speaking, is tastefully appointed—the only items which look at all pseudo or over-styled are the rather heavy-looking instrument panel surround and the single piece of plastic wood around the base of the gearlever. The instruments are very comprehensive—rev counter, speedometer, ammeter, oil pressure, water temperature, fuel and a clock—and really decently executed. The speedometer and tacho are pleasingly big-faced and all figures are plain white on black dials which are angled towards the driver. No vague squiggles to identify vaguer blobs placed poorly in relation to the needle. It is still very recognisably GM with the typical reversed-angle facia but unlike, for instance its Luton cousin the Vauxhall Victor, makes use af all possible locker-space—there's one of those quite practical drop-in pockets in front of the passenger.

The steering wheel is a dished wood-rimmed affair on the end of an energy-absorbing column. Steering is light, accurate, has plenty of feel and seems to be everything it should be. The pedals are pendant and one sits with legs angled outboard noticeably to clear the big transmission tunnel. This is partially due to the manufacturer's wise insistence on mounting the engine well back to prevent the car being too front-heavy with the bigger unit.

The gearchange is delightful—light, precise and short of movement. A lift-up collar keeps the careless out of reverse. Just by the gear-lever and its big rubber gaiter (a "sporty" addition which delights some American's buyers) is the lever which swivels the headlamps when you want to ogle someone.

Seats are unusually high-backed and pretty comfortable; the backs have limited rake adjustment. There is enough leg room for a six-footer and equally good headroom, though one is conscious of the marked tumblehome of the barrel sides. Unlike some other American-owned European manufacturers who make distantly similar sorts of cars, Opel make no dishonest claims about rear room—you *can*, at a pinch get a third person in the back but they wouldn't be too comfortable over a long way. The spare wheel has to be taken out via the cockpit and presumably you're meant to get some of the claimed 5.5 cu.ft. of luggage into the tail.

More searching impressions were prevented by the short time we had with the car. The engine didn't sound particularly sporting—in other words it wasn't noisy, though we were conscious of some tappety clicks. Road noise levels seemed low and it was nice to find a two-speed heater motor that was quiet on full blow. There was a little transmission whine. Visibility forward and directly to the side is good but the fashionably hefty rear three-quarter panels do detract from vision at awkward tee-junctions and in traffic—strange that after all the near-pillarless saloons of the past so many manufacturers should go back towards piggy-windowed bodies.

You can buy an Opel GT now in 1.9-litre form for around £1,800—a trickle is beginning to enter the UK. It is a car that for the moment must appeal mainly to the man who wants something out of the ordinary. It is reportedly selling well in Germany, understandably so, and will undoubtedly appeal a lot in Britain when right-hand drive versions are provided.

ROAD TEST

OPEL GT

"MINI-BRUTE STAGE 11"

Because we've usually driven most of the late-model Detroit products by this time every year, we find ourselves running out of new and exciting autombiles to evaluate. This year, however, a completely fresh and radically different little car has appeared on the scene. Called the Opel GT, it is billed by Buick, who sells the car in the U.S., as "an automobile to light your fire."

Light our fire it did, along with every other person who set eyes on the sleek, razor-nose GT coupe. Even with the minimum of fanfare this Opel has received, we were truly amazed by the number of people who recognized our test car, one of the first to arrive from Germany. The Opel GT became the topic of conversation wherever we went.

There is no need to mention what domestic sports car the Opel has been patterned after. Even with a "familiar" interior and exterior design, the Opel GT seems to have been styled from the ground up, with no other car in mind. Everthing on the car is functional. For example, the lump on the hood was needed to clear the air cleaner, the two rectangular slits above the front grille also serve as cooling inlets, and the rear spoiler does serve its purpose.

The Opel GT is available with two different engine sizes. The standard powerplant has a displacement of 1100 cc's. It develops 67 hp at 6000 rpm with a compression ration of 9.2:1. For $99 more a prospective buyer can get the 1900cc engine, which develops 102 hp. Considering the low additonal cost of the larger, optional engine, we think it is the better choice.

Only one accessory is listed for the GT, an electrically heated rear window for $19. Flow-through ventilation, pushbotton AM radio, radial tires, electric clock, lockable gas cap and power-assisted disc brakes are all standard. A three-speed automatic transmission can be ordered at additional cost for the 1900cc version only.

The basic drive-train on the Opel GT is virtually identical to that of the Opel Rallye. Many other components from Opel sedans were also used to keep the price down.

The handling and driving position of the GT are almost too good to be true. The seats afford good lateral support, and one sits in them enough so he isn't jounced out when cornering hard. There are no separate head restraints since the seat backs extend high enough to guard against whiplash. The steering wheel is small, and the pedals are offset to the right. The seat had plenty of travel, facilitating an arms-out driving posture. It takes a little while before one becomes completely used to the pedal arrangement, but this is nothing unusual with foreign cars.

We had imagined that a car with an optional, heavier engine would accordingly have plenty of understeer. We were pleasently surprised to find that Opel did not follow this tradition. Our 1900cc test car had some body lean, yet the overall cornering attitude was almost neutral. Power could be applied so the rear end came out, but the Uniroyal radials kept things under control.

To get maximum performance out of the coupe, smooth shifts were mandatory. We found that when we tried to bang the lever into gear, the car bucked, then gas surged in the lines so that the carburetor didn't get enough fuel to keep functioning without creating an engine miss. Clutch action was very good, as was the transmission linkage. The short stick was mounted in the right place on the console, providing effortless shifting. From the driver's seat, visibility is excellent. The large tachometer and 150 mph speedometer are placed directly in front of the driver, with the other instruments angled toward him in the center of the dashboard. In front, and off to the left of the shift lever is another lever that at first glance looks like the hand brake. By pushing this lever forward, the concealed headlights revolve around until they are in the night driving position. As soon as they are locked into position, the lights automatically go on. Should the headlights not be opened or closed all the way, a white light will appear on the dash.

The dimmer switch is located on the end of the turn signal stalk.

Braking has to be one of the Opel's best points. No matter how fast the car is driven, it always stops quickly and in a straight line. We did not encounter any fade in the disc/drum combination, and there was only slight nose-dive. The Opel had one of the best braking systems we have ever used.

When a car is in the grand touring category, it usually has some amount of trunk space. Well, the only luggage space in the Opel was behind the front seats, and there weren't even any straps to tie suitcases and belongings down with.

Straightline performance of the Opel was surprising. It went right up to 90 mph in third gear without any complaints or unnecessary fuss. In fourth there is more than adequate passing power, and the engine is fairly quiet. Through the gears we got some buzzing out of the sohc engine, but it was not annoying. Also, the throttle was too sensitive for us. It took a long time to get used to its light springy feel.

After our enjoyable experiences with the Opel GT, one of the finest sports cars to come from General Motors, we can understand why Buick dealers are having to backlog orders. Delivery time is currently several months away, which lends support to Buick's earlier claim that the car is being introduced in limited quantities. Apparently, Buick intends to maintain control over that "fire."

Some body lean, but handling stays pretty much neutral as GT is whipped around slalom course.

Headlights rotate easily when a lever inside the cockpit is pushed forward.

GT-1 Interior is functionally laid out with all necessary controls and gauges close to the driver.

Engine is mounted way back from the nose. Accessibility is tight with 1900cc powerplant.

Test Car: OPEL GT.

PRICE
Basic list: $3395.00
As tested: $3494.26
Options included: .... 1900 cc engine

ENGINE
Type: .... Sohc in-line four cylinder
Bore x stroke: .... 3.66 x 2.75
Displacement: .... 115.8 cu. in. (1900cc)
Compression ratio: 9.0:1
Rated bhp @ rpm: .... 102 @ 5200
Rated torque @ rpm: .... 121 @ 3600
Induction system: .... One Solex 2-barrel
Electrical system: 12v   25 amp alternator
Type fuel required: .... Premium

DRIVE TRAIN
Transmission type: .... Four-speed all synchromesh
Clutch diameter: .... 8.0 in.
Gear ratios: 1st 3.43:1
              2nd 2.16:1
              3rd 1.37:1
              4th 1.00:1
Shift lever location: .... Console
Differential type: .... Hypoid semi-floating
    Axle ratio: 3.44:1

CHASSIS and SUSPENSION
Frame type: .... Unitized
Brake type: ...... Disc-drum combination, power assisted
Brake swept area: .... 277 sq. in.
Steering type: .... Rack and Pinion
    Turns lock to lock: 2.75
    Turning circle: .... 35.4 ft.
Front suspension: Independent with transverse leaf shock absorbers
Rear suspension: Live axle, coil springs, shock absorbers, trailing arms

DIMENSIONS and CAPACITIES
Wheelbase: .... 95.7 in.
Front track: .... 49.4 in.
Overall length: .... 161.9 in.
Overall height: .... 48.2 in.
Overall width: .... 62.2 in.
Curb weight: .... 2072 lbs.
Weight distribution (without driver): .... N.A.

Ground clearance: .... 5.1 in.
Front seat hip room: .... 46.7 in.
    Shoulder room: .... 46.5 in.
    Head room: .... 35.0 in.
    Pedal-seatback, max: .... 43.5 in.
No. of passengers: 2
Luggage space: .... (see text)
Crankcase: .... 2.9 qts.
Cooling system: .... 6 qts.
Gas tank: .... 13.2 gal.

WHEELS and TIRES
Wheel size and type: .... 5J x 13 stamped steel
    Optional size: .... none
Tire size and type: 165 HR x 13 Uniroyal Rally "T" Radial
Normal inflation pressures ...... Front: 22psi—Rear: 24psi
Maximum load per tire: .... 1080 lbs. @ 36 psi

MAINTENANCE
Engine oil change interval: .... 3000 miles/2 months
Oil filter change change interval: .... 24,000 miles
Lubrication interval: .... none

FUEL CONSUMPTION
Test conditions: .... 25.8 mpg
Normal conditions: .... 25-28 mpg
Cruising range: .... 330-370 miles

PERFORMANCE
Zero to: Seconds
    30 mph 3.9
    40 mph 5.8
    50 mph 8.6
    60 mph 11.5
    70 mph 15.0
Standing ¼-mile: .... 17.3 secs.
    Speed at end: .... 75 mph
Top speed: .... 110-115 (est.)
Braking Performance: .... Excellent
    Directional control: .... Excellent
Speed ranges in gears: Rpm's
    1st .... 32mph...6000
    2nd .... 57mph...6000
    3rd .... 90mph...6000
    4th .... 118mph...6000
Rpm redline: .... 6100 rpm

LARRY WILLETT PHOTOS

1.9 OPEL GT

*Not bad but not
up to expectations*

OPEL, SINCE THE 1920s, has seemed to epitomize the dull and uninspiringly Teutonic in automobiles, and GM's takeover of the company in 1929-1931 merely perpetuated this role. But in 1965 a prototype called the Opel GT appeared at the Frankfurt show, seeming to signal a new flash of light at the staid German company. Indeed, the GT was acclaimed the world over, and some of us interpreted it as a harbinger of Corvettes to come. As it turned out, we weren't far wrong on this; its relation to the current Corvette's styling is even more obvious in the production GT than it was in the prototype. The GT, by the way, isn't Opel's first production sports car: before 1914 they were a great power in the sports car world with big, beefy 4-cyl cars that won the 1909 Prince Henry Trials, a joint victory in the 1911 and 1912 Alpine trials, and some other significant honors against rugged opposition like the Porsche-designed Austro-Daimlers, Benzes, Mercedes and Fiats.

The production Opel GT, whose body is being built in France by Brissonneau and Lotz, is essentially a Rallye Kadett living in a new set of clothes. Its wheelbase is 0.6 in. longer and its track a hair wider, but this means only that the suspension units were bolted on at slightly different distance from each other and that the wheels have more offset. Overall, the GT is 2.7 in. shorter, 0.3 in. wider and 7.1 in. lower than the Rallye and seats two less people. Front suspension is unusual, with unequal-length A-arms for the geometry and a transverse multi-leaf spring; rear suspension is what the Kadett got in 1968, a live axle located by simple trailing arms and a Panhard (lateral) rod and sprung by coils. The engine has been moved back approximately 16 in. in the chassis from its Kadett position, to retain the same weight distribution in the face of the large front and small rear body overhangs; the 1.1-liter pushrod 4-cyl unit developing 67 bhp at 6000 rpm is standard, and the 1.9-liter engine, with its unusual high-camshaft, short-pushrod design, is an option costing an additional $99 and developing 102 bhp at 5400 rpm. Disc front brakes come with either engine; the drum rear brakes are larger when the big engine is ordered.

1.9 OPEL GT
AT A GLANCE

Price as tested	$3494
Engine	4 cyl inline, 1897 cc, 102 bhp
Curb weight, lb	2105
Top speed, mph	113
Acceleration, 0–¼ mi, sec	17.7
Average fuel consumption, mpg	27.5

Summary: GM Look comes to a small sports car ... good acceleration & top speed, generous torque ... mediocre handling ... nice brakes ... rather noisy car, but good ride.

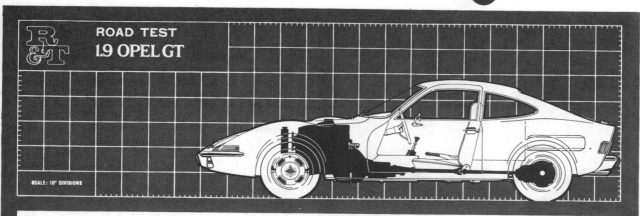

ROAD TEST
1.9 OPEL GT

SCALE: 10" DIVISIONS

PRICE
Basic list.................$3395
As tested...............$3494

ENGINE
Type............4 cyl inline, sohc
Bore x stroke, mm.....93.0 x 69.8
Equivalent in.......3.66 x 2.75
Displacement, cc/cu in...1897/116
Compression ratio..........9.0:1
Bhp @ rpm..........102 @ 5400
Equivalent mph...........106
Torque @ rpm, lb-ft...115 @ 3100
Equivalent mph............61
Carburetion...1 Solex 32 DIDTA-4
Type fuel required.......premium
Emission control......air injection

DRIVE TRAIN
Clutch diameter, in............8.0
Gear ratios: 4th (1.00).....3.44:1
3rd (1.37)..............4.72:1
2nd (2.16).............7.43:1
1st (3.43)............11.80:1
Final drive ratio...........3.44:1

CHASSIS & BODY
Body/frame............unit steel
Brake type: 9.4-in. disc front,
 9.1 x 2.2-in. drum rear; power
 assisted
Swept area, sq in..........277
Wheels........steel disc, 13 x 5J
Tires....Uniroyal Rallye 165 HR-13
Steering type.......rack & pinion
Overall ratio............17.4:1
Turns, lock-to-lock.........3.0
Turning circle, ft..........33.0
Front suspension: unequal-length
 A-arms, transverse leaf spring,
 tube shocks
Rear suspension: live axle on trail-
 ing arms & Panhard rod; coil
 springs, tube shocks

INSTRUMENTATION
Instruments: 150-mph speedo,
 7000-rpm tach, 99,999.9 odo,
 oil press, water temp, ammeter,
 fuel level, clock
Warning lights: oil press, alternator, brake fluid loss, hazard
 flasher, high beam, directional
 signals

ACCOMMODATION
Seating capacity, persons........2
Seat width.............2 x 19.0
Head room..................36.5
Seat back adjustment, deg......15
Driver comfort rating (scale of 100):
Driver 69 in. tall..............100
Driver 72 in. tall...............80
Driver 75 in. tall...............75

MAINTENANCE
Engine oil capacity, qt.........6.3
Every 3000 mi: chg eng oil, chk fluids
Every 6000 mi: chg oil filter, cln air
 filter & fuel filter, tune engine,
 chg plugs, rotate tires, adj clutch,
 chk wheel align
Every 12,000 mi: replace air filter
 element
Warranty, mo/mi......12/12,000
Tire pressures, psi.........24/24

EQUIPMENT
Options on test car: 1.9-liter engine
 ($99)
Other: automatic transmission
 ($190), rear window defroster
 ($19)

GENERAL
Curb weight, lb.............2105
Test weight.................2420
Weight distribution (with
 driver), front/rear,%.....54/46
Wheelbase, in..............95.7
Track, front/rear......49.4/50.6
Overall length..............161.9
Width....................62.2
Height....................48.2
Ground clearance, in.........5.1
Overhang, front/rear...36.5/29.7
Usable trunk space, cu ft......6.6
Fuel tank capacity, gal......13.2

CALCULATED DATA
Lb/hp (test wt)..............23.7
Mph/1000 rpm (4th gear)....19.8
Engine revs/mi (60 mph).....3030
Engine speed @ 70 mph....3560
Piston travel, ft/mi..........1390
Cu ft/ton mi...............83.8
R&T wear index..............42
R&T steering index.........0.990
Brake swept area sq in/ton....229

ROAD TEST RESULTS

ACCELERATION
Time to distance, sec:
0-100 ft..................3.7
0-250 ft..................6.2
0-500 ft..................9.5
0-750 ft.................12.3
0-1000 ft................14.7
0-1320 ft (¼ mi).........17.7
Speed at end of ¼ mi, mph....77
Time to speed, sec:
0-30 mph..................3.6
0-40 mph..................5.5
0-50 mph..................7.4
0-60 mph.................10.8
0-70 mph.................14.7
0-80 mph.................19.1
0-100 mph................34.0
Passing exposure time, sec:
To pass car going 50 mph....6.7

FUEL CONSUMPTION
Normal driving, mpg........27.5
Cruising range, mi..........363

SPEEDS IN GEARS
4th gear (5800 rpm), mph....113
3rd (6100)..................88
2nd (6100).................44
1st (6100).................36

BRAKES
Panic stop from 80 mph:
Deceleration, % g...........81
Control...................good
Fade test: percent of increase in
 pedal effort required to main-
 tain 50%-g deceleration rate in
 six stops from 60 mph......nil
Parking: hold 30% grade......yes
Overall brake rating....very good

SPEEDOMETER ERROR
30 mph indicated.....actual 28.8
40 mph.....................39.0
60 mph.....................59.2
80 mph.....................79.3
100 mph....................99.4
Odometer, 10.0 mi.....actual 10.0

ACCELERATION & COASTING

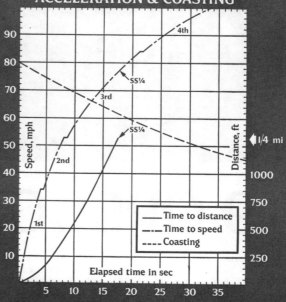

JUNE 1969

Spare tire and tools live behind snap-down vinyl partition.

1.9 OPEL GT

The prestigious Italian quarterly *Style Auto* has awarded its 1969 prize for the "best styled production car" to the Opel GT, but we don't share SA's enthusiasm without reservations. As we've pointed out before, something was lost in the translation from show car to production car—a great deal. In fact, hardly a line remains the same. The styling is swoopy and eye-catching, but the proportions—which were so elegant in the show car—are dinky; lines that look exciting on the full-size Corvette look amusing on the GT. So the GT winds up being the sports-car equivalent of the Kadett —a scaled-down American car that didn't come off.

From a performance standpoint we were prepared for the worst in the GT: the Rallye 1.9 we drove last year had so much engine noise it made us cringe. We were pleasantly surprised here, the GT's engine noise level being similar in character but smaller in volume. Fan roar is still the most prominent characteristic and could be reduced by a viscous fan drive in the GT; but the GT's 1.9 engine, with taller gearing than the Kadett, seems relaxed at its work and has generous torque across its entire speed range. The 1.9 engine also has an automatic choke, a nicety not generally found in 4-cyl imports, and can be ordered with a new 3-speed automatic transmission.

Our test car had the excellent 4-speed gearbox, which we found enjoyable to use in everyday driving and fully up to the job of accomplishing slam shifts in the acceleration runs. Using 5800-rpm shift points we got our GT through the quarter mile in 17.7 sec, quite a bit better than the Kadett's 18.3, and its advantage isn't just at high speeds from the more aerodynamic GT shape; perhaps Opel has done some engine sorting-out in the past year too, even if the power rating is no higher—the GT actually weighs a little *more*. Fuel economy was also surprisingly good at 27.5 mpg; the top speed of 113 mph didn't surprise us. Surely Buick won't sell many of the 1.1-liter GTs in the U.S. now that the 1.9's noise level is within reason and the price premium is so small.

When it comes to steering and handling the GT gets mixed marks. The steering is light and really feels good, though the steering wheel is made of that kind of plastic that gets all slippery when the palms get moist. But the handling of the chassis is at best mediocre. At low speeds the understeer is fierce and in really tight corners the inside rear wheel picks up as easily as that of many Ponycars we've driven. As speed goes up the understeer is somewhat moderated, but even at 50 mph in 2nd gear it's well-nigh impossible to get the rear end out—the front end just keeps plowing. Cornering power is low, no doubt because the 165-13 tires are just too small for a 2000-lb car, and the understeer characteristics must result from using suspension geometry originally designed for a General Motors sedan. For the European market Opel makes available handling options: a rear anti-roll bar, stiffer springs and a limited-slip differential. The first two might do much to reduce the understeer and the third item is a crutch to get around wheel lifting, but none of these is available in the U.S. market.

The GT's ride, however, is quite good. Over tar strips, Bott's Dots and the like the car isn't harsh; over really bad roads the suspension is supple, occasionally bottoming over big dips but not threatening to break the car in two. There is the usual short-wheelbase choppiness, to be sure. The body is commendably rigid and free of rattles, and thus contributes its part to the GT's good rough-road behavior, which is marred only by some tendency to steer in various directions as the wheels hit bumps. On the highway the ride can't be described as quiet, because engine, wind and road noise are all at a rather high level, but it is comfortable.

The disc/drum brakes made a very good impression. They're vacuum assisted but not overly so, they pull a decent deceleration rate in the panic stop with good control (the front ones lock first, and discs are easier to get unlocked than drums) and they pass our fade test with no increase in pedal effort or unseemly pulling.

A body whose shape is so governed by Styling as is the GT's is bound to be inefficient in its use of space; this is no exception. Shoulder room is restricted by the waist pinch, and luggage space if not deficient is difficult to get at be-

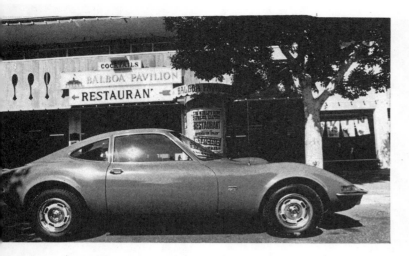

1.9 OPEL GT

cause there's no outside access to it. The driving position is good, the seats well contoured and firm with adjustable backrests, and the instruments and controls laid out within easy reach even with the messy separate shoulder belt fastened. Instrumentation is exceptionally complete: the large and readable speedometer and tachometer are right ahead of the driver but too low and the four minor gauges, as well as a clock, are in the center dash area, but close enough to and leaned toward the driver. The ammeter and oil-pressure gauges have warning lights built into them—the dual system we've long advocated. Pedals are small but widely spaced, and the shift lever and handbrake are just where we like them. The retracting headlights—can't somebody design an aerodynamic body without these bothersome things?—are put up by sheer physical force through a hefty lever in the central console.

High seatbacks, rather than the usual form of adjustable head restraints, give a luxurious feeling but further reduce the already marginal rear vision. To the front things are better, with the ventwings gone in favor of good flow-through ventilation with separate dash and foot vents. Still, opening a side window increases flow through the system—hardly anybody has really learned how to get full flow through an interior with body exits yet. There is no locking glove box, only an open bin on the right that also forms a grab "handle"—but all things considered we feel the GT's interior is rather nice and attractive. And in general the assembly quality and finish of our test car were up to what we expected; Brissonneau and Lotz get especially high marks for the excellent exterior panels.

At $3500 the Opel GT is a unique quantity, certainly worth considering against other sporting cars in the same price range. Its straight-line performance is fine and it has good brakes; the handling may not be good but it is predictable and safe. There is nothing exotic or unfamiliar about its mechanicals, and there's a Buick dealer in nearly every village. Perhaps we'd even be kinder about its styling if Opel hadn't shown us that delectable prototype first.

Opel GT

Buick introduces automobiles

Buick introduces the Opel GT. Who else but Buick would?

Buick introduced you to the Riviera GS. Buick introduced you to the California GS.

Buick brought you the GS 400. Buick brought you the GS 350. Buick brought you the Opel Rallye Kadett.

Buick introduced you to Stage I.

You should have known who was responsible for the Opel GT the minute you saw it. Who else?

Now that Buick has introduced the Opel GT, you have every reason to be excited. Only, please. Have patience.

Because, as much as you may want this automobile, you may not be able to get one right away. Like all good things, the Opel GT is available in a limited quantity.

Which, if you'll think about it, is really the way it should be.

Automobiles like this simply don't come along very often.

Especially automobiles equipped like this that nearly everyone can afford.

There really isn't much you could ask for that isn't already standard on the Opel GT.

A 67 horsepower engine is standard. A 102 horsepower engine is available for the most enthusiastic enthusiast.

YOU MAY PURCHASE A FULL COLOR 18 X 37 INCH POSTER OF THE OPEL GT. SEND A CHECK FOR $1.00 TO BUICK'S OPEL GT/P.O. BOX 5368-C/DETROIT, MICHIGAN 48211.

Riviera GS

to light your fire.

Power assisted front disc brakes are standard.

A short-throw, four-speed stick is standard. Mounted in a console on the floor. A fully automatic transmission is available. Mounted in a console on the floor.

Dual exhausts are standard. Full bucket seats are, too. Specially contoured. Dressed in vinyl.

Radial ply tires (165 HR x 13) are standard.

Instrumentation that includes a tachometer, amp and oil gauges, and 0.1 mile odometer is standard.

And, of course, a very lengthy list of General Motors safety equipment is standard, too.

There you have it. The exciting beginnings of a long story that can be heard in full at any Buick-Opel showroom.

Which is where you'll find Buick's Opel GT.

Where you've been able to find Buick's Riviera GS. Buick's California GS. Buick's GS 400 and GS 350. Buick's Opel Rallye Kadett. And Buick's unique Stage I.

All of them brought to you by you know who.

Who else?

Gentlemen, warm up.

The light-your-fire Buicks: Riviera GS. GS 400. Stage I. GS 350. California GS. Opel Rallye Kadett. Opel GT.

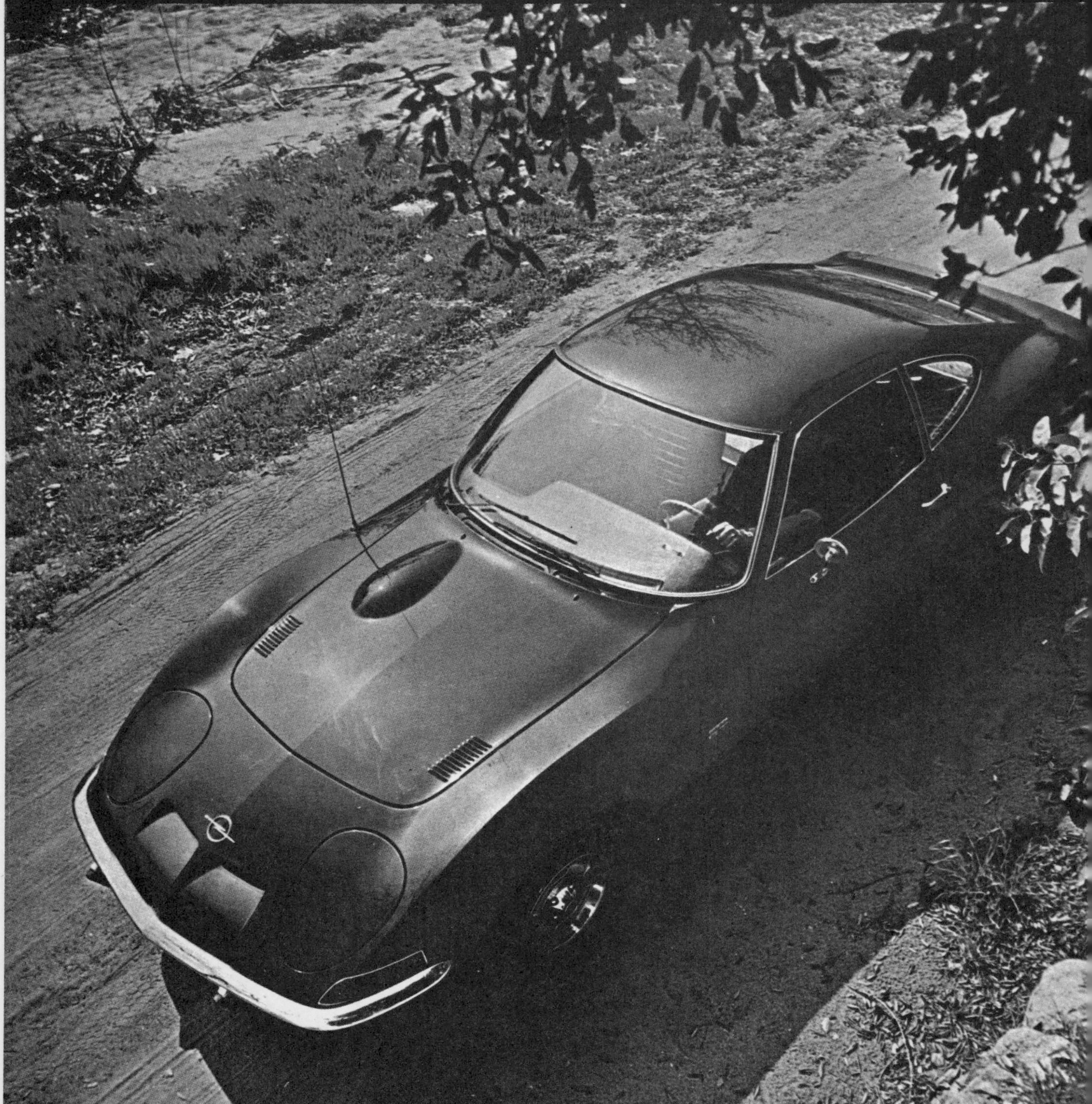

OPEL GT BIG SURPRISE...

. . . at your friendly Buick dealers. It looks like a Mini-Vette, and handles more like a 'Vette than an Opel.

WHAT'S AN OPEL GT? It's the Mini-Brute's Mini-Vette. It's a 1.9-liter sports car from Opel, GM's German subsidiary. It's a low cost, medium-performance GT car sold by America's huge Buick dealer organization. It's GM's bid in the lucrative low-price sports car market currently being dominated by Datsun, Triumph, MG, Fiat, et al.

But to many Americans, it's going to be "Buick's new sports car." More importantly—to its owners—it will be not only a car that will handle as well as the competition in its price range (roughly $3500 loaded), it offers two extras not necessarily available from the others—distinctive style, plus service and parts as near as your local Buick dealer.

The big surprise is how Opel achieved the comfort, convenience, style and engineering. The first Opel GT was shown two years ago as a styling exercise on the Kadett chassis, Opel's economy sedan. The announcement last fall that it would be produced was not terribly earth shattering. Pretty, certainly; but we had our doubts about the worthiness of the en-

PHOTOS BY DAVE GOOLEY

CAR LIFE ROAD TEST

HANDLING was nimble and responsive at moderate speeds, but cornering near the limit of adhesion revealed too much understeer and rear wheel lifting.

OPEL GT
continued

LARGE DOORS and roomy interior put GT a cut above most sports cars. Instrument panel was stylish yet logical; center dials angle toward driver.

SPUNKY 102-bhp engine has been moved back several inches and valve cover chopped as concession to GT hood. Note casual cold air package, front mounted brake booster (upper center).

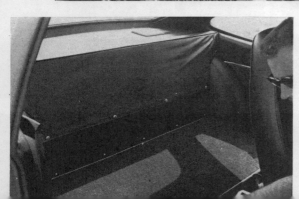

NO trunk, à la Corvette, but luggage area is larger and easier to reach than 'Vette's.

gine and chassis, even though the Rallye Kadett we tested (December, 1968) was impressive in both the performance and handling departments for an economy sedan.

But rebodying the Kadett, we felt, would not make a very sophisticated sports car. To our surprise, Opel has done essentially that (plus detail changes), and has come up with a nice integration of power, handling, size and style. The basic chassis is that of the Kadett. The engine has been moved back roughly 12 in.; the passenger compartment was shifted to the rear a proportional amount; and the smoother, lower body was added. This brings down the center of gravity (the Kadett felt very tippy) and allows the axle ratio to be lowered numerically due to the lowered drag.

Standard engine is the little 66-cid 4-cyl. ohv with 67 bhp. But optional, and the obvious choice of the enthusiast, and the one that will make the GT a genuine sports car, is the modern 102-bhp/116-cid Rallye engine. Our test car, with this engine and the standard four-speed transmission, came off on the specification sheet as a carbon of the Rallye Kadett previously tested. Only the rear axle ratio (3.44:1 vs. 3.67:1 on the sedan), and tire size, are different (165-13 radials on the GT, 155-13 on the sedan).

However, differences not apparent on paper are quickly obvious at the test track. Quarter-mile times are nearly a second quicker and five miles per hour faster. Top speed, 111 mph, is better by 10 mph over the sedan. Braking is slightly better, and handling is vastly improved with the lower center of gravity. We're not exactly sure where this extra performance came from. The engine is *supposed* to be the same. Only readily apparent inducements for quicker acceleration is 100-rpm higher redline (6000 rpm), slightly larger section tires and a very grippy clutch. It could be moved off the line very smartly, with just a whisper of wheelspin. Snap shifts were easy and the clutch and tire grip were

INTERESTING LIGHT SHOW was amusing to watch, noisy to listen to. Mechanical linkage was activated by firm push on console handle. If firm enough, lights would slam home with loud thump. Last portion of movement turned on switch—sometimes.

good enough to send a judder of torque throughout the car. The slippery shape helped on the top end.

We still don't like the transmission ratios. First and second, and third and fourth are closely spaced, leaving a knee-buckling drop between second and third. This not only has its disadvantages in acceleration, but often one is left without a proper gear for certain corners, where the choice is lugging around in third, or buzzing along well past the power peak in second. Thankfully, the GT has been spared the horrendous fan noise of the sedan. The higher rear axle gearing reduces the engine speed for corresponding road speed, and more attention was paid to internal aerodynamics allowing a smaller and slower radiator fan. The noise is still there, more than we would like or are used to in domestic cars; but at least it doesn't drive the passengers out of the car.

The disc/drum brake system, as expected, was top drawer. Best deceleration rates in our stops from 80 mph were excellent, though slightly erratic, between 31 and 28 ft./sec./sec. Fade was not evident in eight consecutive stops. A slight and nearly imperceptible servo-assist is added by a curiously placed vacuum diaphragm. Instead of hanging on the firewall, diaphragm and master cylinder have been moved forward to the radiator bulkhead with a linkage rod spanning the length of the engine compartment to connect pedal to master cylinder. We're not exactly sure why. There seems to be room on the firewall.

We took the opportunity to evaluate the handling of the GT on both a very tight slalom course and on high-speed mountain roads, all on the same day. Bob Bondurant's School of High Performance driving was operating at Orange County Raceway on test day

ANY DOUBT about its ancestry? Though not an exact replica, the GT does borrow several styling details from the Corvette, especially the nose and tail treatments. It got more curious stares than the Improbables.

OPEL has its moment on competition driving school course. GT could stay with novices in prepared Datsuns, but had to be hung out to do it.

and we got permission from the instructor to lap the tight course with his students in Datsun 2000s. After a few laps we began to get the feel of it and we (a) surprised even ourselves, (b) watched the instructor's sneaky grin turn into a frown when we started catching his students, and (c) found the GT's two worst faults.

Later, in the mountains, we reaffirmed what we suspected on the test track: Heavy initial understeer and rear wheel lifting. For easy "fun" driving around the neighborhood, the car seems nimble and reasonably responsive, but out at the limit, the car feels unpredictable and lacks real cornering power.

Two things: It needs a rear antiroll bar to take out some of the understeer, and a limited-slip differential to get the power to the ground coming out of a corner. So equipped, the GT would probably be as impressive a handler as one can buy short of a more expensive Corvette or Porsche.

Interestingly, the Rallye came standard with the rear bar, and first GT press releases last fall listed a handling kit of stiff springs and shocks, rear bar, and a limited-slip differential. Later releases neglect to mention this, but show a sketch of the rear axle

1969 OPEL GT

CHASSIS/SUSPENSION
Frame type: Unitized.
Front suspension type: Short and long arms transverse.
 ride rate at wheel, lb./in. . . . . . . n.a.
 antiroll bar dia., in. . . . . none
Rear suspension type: Live axle, coil spring, torque control arms, track bar.
 ride rate at wheel, lb./in. . . . . . . n.a.
Steering system: Rack and pinion.
 overall ratio 17.4
 turns, lock to lock 3
 turning circle, ft. curb-curb 33
Curb weight, lb 2070
Test weight 2380
Test weight distribution, % f/r . 55/45

BRAKES
Type: Power-assisted disc/drum.
Front rotor, dia., in. 9.4
Rear drum, dia. x width . . . . . 9.0 x 2.2
 total swept area, sq. in. 277

WHEELS/TIRES
Wheel rim size . . . . 13 x 5J
 optional size . . . . none
 bolt no./circle dia. in. 4/4
Tires: UniRoyal Radials
 size 165 HR-13

ENGINE
Type, no. of cyl. IL-4
Bore x stroke, in. . . . . 3.66 x 2.75
Displacement, cu. in. 115.8
Compression ratio 9.0:1
Fuel required . . . . premium
Rated bhp @ rpm . . . . . . 102 @ 5400
 equivalent mph 104
Rated torque @ rpm . . . . 115 @ 3100
 equivalent mph 60
Carburetion: Solex 1x2.
 throttle dia., pri./sec. 0.95/1.1
Valve train: Overhead cam, tappets and rocker arms.
 cam timing
 deg., int./exh. 44-86/84-46
 duration, int./exh. 310/310
Exhaust system: 4 into 1 manifold, reverse flow muffler, transverse resonator with dual tail pipes.
 pipe dia., exh./tail 1.5/1.2
Normal oil press. @ rpm . 37 @ 3000
Electrical supply, V 12
Battery, plates/amp. hr. n.a./44

DRIVE TRAIN
Clutch type: Single dry plate disc.
 dia., in. 8.0
Transmission type: Four-speed manual, fully synchronized.
Gear ratio 4th (1.00:1) overall. 3.44:1
 3rd (1.37:1) 4.71:1
 2nd (2.16:1) 7.34:1
 1st (3.43:1) 11.8:1
Shift lever location: console.
Differential type: Hypoid.
 axle ratio 3.44:1

DIMENSIONS
Wheelbase, in. 95.1
Track, f/r, in. 49/50
Overall length, in. 162
 width 62
 height 48
Front seat hip room, in. . . . . . 2 x 21
 shoulder room 50
 head room 37
 pedal-seatback, max. 43
Door opening width, in. 43

PRICES
List, FOB New York $3395
Equipped as tested $3513
Options included: 1.9S Engine . . . $99
 rear window defroster $19

CAPACITIES
No. of passengers 2
Luggage space, cu. ft. 7
Fuel tank, gal. 10.5
Crankcase, qt. 3
Transmission/dif., pt. 2.5/2.5
Radiator coolant, qt. 6

DISAPPOINTING weigh-in revealed a slightly heavier curb weight and worse distribution than sedan.

with and without a rear bar. Nor do the option lists contain the handling kit or any part of it. At this writing, the kit is probably for European sale only and will be offered here when and if the demand is great enough. The sedan handling package by the way, has both items, and will bolt right on.

Inside, the Opel has one of the better people packages of the sports and GT car world. The larger members of our staff found entrance and egress better than the Corvette and most imports. The large door, which opens high (for once the stylist lost to the comfort engineers), wide (some 43 inches, close to a full-size sedan) and handsome (stylist didn't lose all the battles). Door location relative to the seat is also good. The large seats are comfortable and pretty fair buckets.

The rest of the interior is plush, logical and roomy. Dash has the tach and speedometer where it counts, and easily read engine instruments on the center section.

Only rub is the international symbology on the rocker switches. Took a long time for us to figure out the one with the jiggly lines on it. It was for the rear window defroster, which was heated by (you guessed it) a bunch of jiggly wires. ■

CAR LIFE ROAD TEST

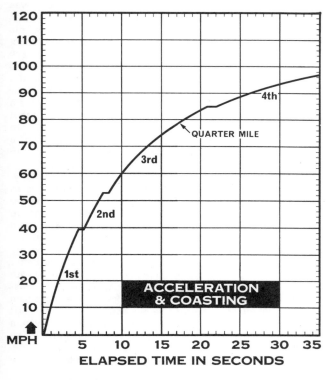

CALCULATED DATA
Lb./bhp (test weight)..........23.4
Cu. ft./ton mile...............119.2
Mph/1000 rpm (high gear).....19.32
Engine revs/mile (60 mph).....3100
Piston travel, ft./mile.........1420
CAR LIFE wear index...........44.0

SPEEDOMETER ERROR
Indicated	Actual
30 mph	28.0
40 mph	38.3
50 mph	48.7
60 mph	59.0
70 mph	69.8
80 mph	78.2
90 mph	87.0

MAINTENANCE
Engine oil, miles/days......3000/60
  oil filter, miles/days.....6000/120
Chassis lubrication, miles......none
Antismog servicing, type/miles
  clean/6000; replace/12,000
Air cleaner, miles.........clean/6000
Spark plugs: AC 42FS.
  gap, (in.)....................0.030
Basic timing, deg./rpm..........n.a.
  max. cent. adv., deg./rpm.34/3200
  max. vac. adv., deg./in. Hg..21/13
Ignition point gap, in..........0.018
  cam dwell angle, deg...........50
  arm tension, oz................17
Tappet clearance, int./exh.
  ....0.012/0.012
Fuel pressure at idle, psi.......2.5
Radiator cap relief press., psi...8.0

PERFORMANCE
Top speed (5800), mph..........111
Test shift points (rpm) @ mph
  3rd to 4th (6000)..............85
  2nd to 3rd (6000)..............53
  1st to 2nd (6000)..............39

ACCELERATION
0-30 mph, sec.................3.3
0-40 mph......................5.1
0-50 mph......................7.4
0-60 mph.....................10.2
0-70 mph.....................13.7
0-80 mph.....................18.0
0-90 mph.....................26.6
0-100 mph....................39.6
Standing ¼-mile, sec.........17.4
  speed at end, mph...........79.0
Passing, 30-70 mph, sec......104

BRAKING
Max. deceleration rate from 80 mph
  ft./sec./sec..................31
No. of stops from 80 mph (60-sec. intervals) before 20% loss in deceleration rate .... 8—13% loss
Control loss? None.
Overall brake performance..excellent

FUEL CONSUMPTION
Test conditions, mpg..........17.6
Normal cond., mpg............22-26
Cruising range, miles.......230-270

WHY THE EXCITEMENT AT

Opel GT a new, honest to goodness sports car for highway or race course.

LOS ANGELES is the most mobile city on earth, and the local motorists are quite hardened to seeing unusual cars sailing along the freeways. You know you have a real winner when people stare at your car while trying to survive in the frenzy of L.A. traffic. This is precisely what happened to us when we took to the road with a new Opel GT. One of the curious even followed us down an off ramp to shout the question, "What is it?" Despite the rather prominent Opel sign on the tail, the questions continued — in the gas stations, parking lots, every place we stopped. Almost everyone guessed it to be some sort of Italian sports car, and most were amazed when we proclaimed it to be from General Motors of Germany.

Opel cars have made giant inroads on the imported car market in this country in the last few years. Well planned merchandising through selected Buick dealers has put the line near the top of the list on import sales and it keeps climbing. The basic sedan and station wagon are the big items, and Opel went for the performance image with their Rallye Coupe, which is the standard body with fancy paint, performance engine, driving lights, lots of instruments and radial tires with special wheels. But a true high performance car, designed as a GT from the ground up, was in development when Opel first entered the U.S. market in a big way. At the New York Auto Show in the spring of 1966 Buick displayed an experimental sport coupe from Opel which bears a marked resemblance to the car we tested.

Buick planned to introduce the car to the American public last winter, but various production hangups delayed its arrival on these shores. One of the biggest obstacles was the production of the all steel bodies. Built in France by Brissoneau and Lotz of Paris the bodies were delayed by the general strike in that country last summer and the whole program was put back about six months. We hear rumors also that certain small problems on U.S. certification were present in the pilot models, but the car was made street legal and is now for sale in this country. It is probably back ordered at most dealerships.

At first glance the most common impression of the Opel GT is that it does have a family resemblance to what might be called its big brother — Chevrolet's Corvette. Then you might think that it looks somewhat like the Toyota 2000 GT or the Alfa Romeo GTZ. Actually it is an individual design in the current vogue with a definite Kamm effect tailbone and the long, low sloping nose to accommodate the conventional front engine/rear drive configuration. The body was wind tunnel tested before the design was finalized and the sleek aerodynamic shape is efficient. A pleasing by-product of the body design is the nearly total lack of wind noise at speed. The clean body is devoid of chrome do-dads and the plethora of stripes that so many manufacturers deem necessary to announce a high performance car, and we like it that way. Even the engine hood louvers are real and functional! Decor consists of a small badge on the front fenders and an emblem on both the nose and the tail.

The front bumper seems overly husky, but no doubt is just the proper size to meet Federal regs. At the rear the two bumperettes enhance the go-fast look to the chopped tail style and the oversize tail lights are properly round and purposeful. The smooth nose houses the pop-up headlights, and these look better when open than most on cars fitted with hidden headlights. Actuated by an unassisted mechanical lever in the cockpit, on the console, the lights swing open 180° and close when the lever is given a healthy shove toward the firewall. The light switch goes on at the same time. The amber front turn indicator lights are

ROAD TEST

The padded dashboard holds the no-nonsense tach and speedo, auxiliary gauges are in the center above the radio, accessory switches, and the heater controls.

BUICK-OPEL DEALERS

mounted in the underslung flat black grill and the high beam switch is located at the end of the turn indicator stalk to the left of the steering wheel. By the way there is room in the grill for a set of driving lights, and there is a switch for them on the center of the dash; actually the European cars have the auxiliary lights installed. Our guess is that Opel had such a headache with various state laws when they put driving lights on the Rallye Coupe that they intend to let the domestic GT owner make his own choice of auxiliary equipment.

The sporty new Opel is actually based on the Kadett chassis platform, but the front end is considerably changed from the bread and butter car, because the engine is mounted several inches further behind the front wheels on the GT. The steel body and chassis comes off feeling quite rigid and strong. In typical Germanic fashion the doors are quite solid and close easily, quietly, and firmly. The overall quality control is high and some of the detail work is equal to that found on much higher priced machines.

Power train

The heart of any car is the engine, and one with such racy looks should have performance to match. The standard engine for the Opel GT is the 1.1 liter (1100 cc) four-cylinder, in-line, water-cooled unit that powers the base line models. However, we doubt if many, if any, GTs will come into this country with the small engine, which is only one hundred dollars cheaper than the 1.9. The test car's power plant is the most likely for domestic sales.

First seen in this country in the Rallye Coupe the 1.9 liter Opel engine is somewhat unique in design. Displacing 115.8 cubic inches (1897 cc) this four-cylinder, in-line, water-cooled engine lies somewhere between a true overhead cam layout and the more conventional overhead valve style from Europe. The camshaft is actually in the head over the wedge shaped combustion chamber, but under the rocker arms. The valves operate without pushrods, but hollow tappets are interposed between the cam and the rocker arms. Drive is by a hydraulically tensioned duplex roller chain. This somewhat unusual valve gear arrangement keeps the engine height down from the norm. Induction is by a single, two-throat, downdraft Solex carburetor that incorporates an automatic choke. With a bore of 3.66 inches and a stroke of 2.75 inches, the 1.9 engine puts out 102 de-smogged horsepower at 5200 rpm and has a torque rating of 121 lb./ft. at 3600 rpm. The 9.0 to 1 compression ratio doesn't seem high by today's standards, but premium fuel is a must for this Opel. One fault with the engine is high noise level, even at mid-range on the rpm scale. The GT is quieter than the 1.9 Rallye Coupe, but the fan racket and general engine roar is still quite noticeable at normal cruising speeds.

The exhaust system utilizes twin mufflers and dual pipes, but it fails to make the expected sporting sounds. The enthusiastic owner might go for different mufflers to get the zippy noise that would seem right for the car. In fact, there is plenty of room in the engine bay for a new manifold with dual carbs and the rest, and we speculate that this operation would raise the horsepower by a bunch, but probably wouldn't pass the federal smog sniffers acid tests. Some of the roominess in the engine bay is accomplished by placing the battery for the 12-volt system forward of the radiator on the floor of the nose piece on the right side, and the power brake booster is placed on the forward crossmember on the opposite side.

The lever for the standard four-speed, all-synchro gearbox is mounted on the console in the center of the car. There is an optional three-speed automatic transmission, but so far the manual is the only model available for test. The gear shift lever is well positioned for most drivers and it has the sporting short throw. The

JULY 1969

The aerodynamic lines are not spoiled when the pop-up headlights are swung open. The dog-bone treaded radial tires perform every bit as well as they look.

The storage space behind the seats is limited but adequate. The button down vinyl flap opens to reveal the spare tire and tool kit on the shelf over the gas tank and under the rear window.

pattern is the standard H and ratios are: 1st — 3.43, 2nd — 2.156, 3rd — 1.366, 4th — 1.000, and reverse — 3.317. The synchros are hard to beat and the linkage is smoothly Teutonic in operation. There is a lift-up collar on the shift lever that locks out reverse gear, and the pattern is marked on the lever knob. The clutch is a single dry plate, diaphragm spring type with an eight inch diameter and mechanical actuation. The hypoid type differential on the rear live axle has a final drive ratio of 3.44 to 1.

Roadability and handling

In a small engined sports car the handling makes or breaks it on the market. The relatively small European engines are not noted for blistering straight line acceleration, but the cars gain their faithful following from the sheer roadability and the maximum use of the available power. The Opel GT is one of these fun cars to hang your hoof into, and the roadability is superb. The GT has none of the wind wander and general twitchiness so common to the more sedate Opels, and the GT is extremely stable in cross winds and in over-enthusiastic cornering situations. Handling tends toward slight initial understeer, but as cornering speeds increase the car becomes neutral and stays positive. We could hardly break the rear end loose and the handling made us look like a real hero. On wet pavement it is easier to hang the tail out a bit, but the car comes right back from the ragged edge to stability most easily, with proper use of the throttle. Even the most timid can flat'fly through the mountains with this very forgiving car. The nifty radial tires contribute a good deal to the excellent handling and traction. Made in Germany by UniRoyal, the nylon, tube type radials are called Rallye "T" and the 165 Hr 13 size tire is mounted on 5J x 13 steel wheels.

Cruising down the highway, legal speeds are reached quickly and it is tempting to think about burying the foot a bit more into the throttle. The lack of wind noise and the sure-footed characteristics of the GT make you wonder if the speedometer is right on. The speedo in our test car was quite accurate with less than one per cent error. A handy touch is the near synchronization of the tach and speedo. At 65 mph the needles are pointing nearly straight up with the tach reading 3500 in fourth gear.

But the twisty side roads are the places to drive the GT. We startled some of the regulars on our favorite stretch of the infamous Mulholland Highway high in the mountains in the city of Los Angeles. The saying "it sticks like glue" is apt for the Opel and there is a certain snob appeal in blowing off more powerful and expensive equipment on a tight mountain road. The gear spacing is good for enthusiastic driving, and second and third do nicely for flogging around the curves. Smooth upshifts and downshifts are easy with the handy and positive gearbox. Steering is quick and precise. The rack and pinion steering unit provides three turns lock to lock, and the small wood-rimmed steering wheel is well positioned for any size driver. The deep dished three spoke design is handy for hooking the thumbs and straight arm driving still leaves the pedals and gear shift within easy reach for all but the under five-footer.

Power and performance

Moving away from the mountains toward the harsh reality of performance testing at Orange County International Raceway, we lined up before the drag strip's Christmas tree and took off. A creditable quarter mile time of 17.82 seconds and a terminal speed of 76.46 came up on several runs. The tires spun just a bit off the line and then possibly they exhibited too much grip for this type of test. From a standing start we reached 30 mph in 3.9 seconds, 40 in 5.3, 50 in 7.6, and 60 in 10.9. We were shifting right on 5000 rpm, although the redline is at 6100 rpm and the orange zone at

ROAD TEST

The half bumpers in the rear enhance the sleek look of the Kamm effect tail design. The large tail lights look good and are truly visible in American traffic.

5300; we found that just over five grand brought out the best in the engine. We ran out of road before peaking out in top gear, but we estimate top speed of the Opel around 110 mph. All the top speeds recorded were reached without getting into fourth gear, although normal shift points would be considerably lower in rpm. Actually the engine will pull easily from 2000 rpm in fourth gear and the middle range torque is quite strong.

The road course at the raceway has some really tight turns on the back section. Running the standard tire pressures (22 psi all around) we provoked lots of tire squeal at 40 mph on the sharp right-hander, but pumping just a few more psi in the front, and a bit more yet in the rear, the noise abated and the turn could be taken much faster on the same line. It is difficult to get a wheel to lift in the corners and the GT must be tossed about violently before it will get out of shape. With relatively conventional suspension the Opel engineers have found excellent balance for the GT and the joy of driving this car hard puts one in mind of his first experience in a real sports car — which was probably an MG and probably remembered as being flawless through the mists of nostalgia.

The GT's suspension is a carryover from the Kadett models with the independent front suspension incorporating double A arms and transverse leaf springs with tubular shocks. At the rear the live axle set-up couples with coil springs, gas-filled tubular shocks, torque arms and a single transverse traction bar. Put together with a fairly low center of gravity it all makes for a delightful package on the road. Weight distribution is good with 1155 pounds (55.2%) at the front and 945 pounds on the rear. The weight as tested with half a tank of gas is 2090 pounds.

Fuel economy is good but not super for this size engine. Bear in mind, however, that during the two weeks spent on test, any car with sporting instincts is driven to the fullest by a number of people and the resulting fuel figures are seldom near optimum. We recorded as low as 22 mpg and up to 28 mpg on the test car, and with normal driving habits, but with full use of the gears, an expected average would be around 25 mpg. The thirteen gallon tank supplies well over a 250 mile driving range, and any car that will take you from LA to Las Vegas without stopping, and be so enjoyable to drive at the same time, more than fits its title of a Grand Touring machine.

Brakes and safety

The disc front, drum rear brake set-up is common on European performance cars and is becoming standard on some domestics. The GT has excellent brakes with the 9.4 inch front discs giving an effective braking under all conditions and the rear 9.1 inch diameter drums supplying a good balance. The mechanical hand brake works off the rear drums from a British type flyoff handle mounted on the rear of the center console. It shouldn't be difficult to put good brakes on a one ton car, and those on the Opel are really fine with an ATE power boost system that is barely detectable to the driver. In fact the pedal pressure is as firm as it should be on a sports car. We had to look under the hood to be sure the booster was really there. We ran a series of stops in rapid succession from 60 miles an hour and were not able to induce noticeable fade, and average deceleration came out at 27 ft. per sec^2 stopping from 60 mph in 142 ft. In fact we couldn't even detect the tell-tale odor of frying linings often encountered in this type of test. The Opel stopped straight and true every time, with just a suggestion of rear wheel lock-up and gentle slewing at the finish of a real panic stop attempted when the brakes were cold. Of course, a brake test of this kind is really half tire test, and the standard radials are a big help in good adhesion on fast stops. In general the brakes are responsive to light pressures and stay positive in the wet as well as in the dry. Tire adhesion, of

JULY 1969

In hard corners the little import behaves like a good small sports car should. It is almost inevitable that the Opel GT will be seen in SCCA racing soon.

course, is less perfect on wet pavement, but the side cuts on the UniRoyals make for an excellent rain tire.

All the required safety items are on the GT and then some. The safety belts are standard GM as are the shoulder restraints. The belts are mounted a bit too high over one's waistline, and the shoulder harness fits at an awkward angle. The belts are so installed that the buckle is only a few inches from the floor on the door side of the car, and the easiest way to buckle up is to do so before the door is closed. However, most owners will probably want to tailor these restraints to fit their own needs. Going along with the rules is the deeply padded dashboard with recessed instruments, the brake warning light, flasher unit, side lights on the exterior, and our particular pet peeve on new cars, a steering lock device. Basically a sound ideal the steering lock is often more an annoyance than a saving grace, and many times on domestic cars we have left the keys dangling in the ignition from the sheer frustration of a hangup in the release device. Our test Opel had the opposite problem. The ignition key slot seems innocently mounted on the right side of the steering column, but as often as not it took several tries at insertion before the proper meshing occurred and the ignition could be activated. But with usage we discovered just the right spot for the key and the problem evaporated. Happily, unlike its domestic built cousins, the Opel will fire in neutral or in any gear with the clutch depressed.

Another lock that can be defeating is the gas cap device. Swinging the emblem sideways reveals the locking gas cap head. Pop the key in and lift up you might think. Not really. The key releases the lock OK, but there is a fiendish trick to opening the cap that calls for just the right side slip on the angle of attack before the cap will pop open. It does make quite a conversation item in the gas stations, however, and you can be assured that nobody will ever be able to steal your gas cap.

Comfort and convenience

Although it looks very small next to most cars, the Opel GT is roomy inside. There is no place for a back-seat and no attempt at a 2 plus 2. Rather it is a real Grand Tourer with really neat bucket seats. They seem to fit everyone who tries the car. The seats are handsome in appearance with well fitting headrests, and they provide good support for the back and the legs. The recliner mechanism and the full travel track offer a myriad of seating positions from the full layback European style to the bolt upright over the wheel position favored in some circles of driving. The pedals are slightly offset, but well spaced. The gas pedal is positioned in good relation to the brake if one cares for the heal and toe style of driving. The small steering wheel is not adjustable, but is mounted fairly low, and the seating position options make for comfortable driving for anyone. The horn button is in the center of the steering post, and we particularly appreciate this, since many current cars craftily conceal the horn so that the unfamilar driver may never find it.

The full panel of instruments by VDO are black face with white markings and they are easily visible at a glance. The large, round tachometer and speedometer are mounted to the left and right of the steering column respectively, and are right in view without conscious effort. The speedo has a total mileage counter with a tenths increment, but sadly there is no trip odometer for this or any other Opel. At the top of the dash in the center are three smaller dials: from left to right, the first reads amperage and oil pressure, the next dial has the water temperature and gas gauges, and the third is a very accurate time of day clock. Under this is the Delco push-button radio that is standard equipment, and under the radio is a brace of buttons with the usual Continental pictures depicting their purpose. From the left, the first button is a

ROAD TEST

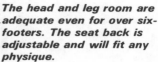

The head and leg room are adequate even for over six-footers. The seat back is adjustable and will fit any physique.

Foot room is generous and the pedals well spaced for a foreign built car. The brake and gas pedals are ideally situated for heel and toe style driving.

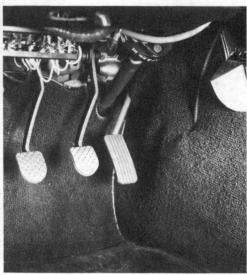

light switch for the non-existent driving lights, the next is a total blank, and the third turns on the rheostat on the dash lights. In the center is the cigarette lighter, and then comes the windshield wiper switch for the two-speed wipers, next the button for the rear window defroster, and finally the switch for the fan on the heater-fresh air system. The heater controls are just below all these switches and the heater puts out more than enough warmth for comfortable motoring in any weather. Under the heater, and nearly flush with the console is the brake warning light and flasher switch. Under the dash by the speedo is the pull lever to release the engine hood and at the right on top of the dash is an open-topped cubby hole that passes for the glove compartment. Fresh air vents are on the lower dash on each side and air is exhausted through vents above the rear window. Ventilation is quite good for a car with the one piece window design. At the extreme left on the console is the lever for the headlights. Unfortunately this console is covered with what looks like wood-grained contact paper and on the test car it had air bubbles in it and looked cheap. This is completely out of character with the quality of the rest of the car.

The instruments are easy to read at night, but the row of switches is not illuminated, and one must become familiar with them to find the right one in the dark. However, in just a few miles everything seems to fall into place and the car fits as if it had been individually tailored for the driver. A day and night mirror hangs from the roof in the center of the cockpit, and this suffers from some vibration at speed. The door mounted mirror on the driver's side is steady as a rock, however, and very handy. Behind the passenger's head is the interior light that is activated by opening the door or by a hand switch on the light itself. The rear luggage compartment behind the seats is the entire storage area. There is no trunk at all. Two small suitcases and various well-planned oddments will fit nicely behind the seats, and just below the rear window is a curtained off shelf that houses the spare tire and tool kit. This is accessable only from the interior and it could make a mess if one had to change a tire in inclement weather, but there is literally no place else to put the spare.

Entry and exit is very good for the four-foot high car. The doors are cut slightly into the roof to aid the art of entry. The doors open wide as well, and there are well-placed arm rests fitted into each door. The opening handles are in an easy spot to reach when one leaves the car, but the window winders are placed low and forward on the door and they are awkward to operate while on the move.

Weather proofing was excellent in the rain storm during our test, but we would strongly suggest getting the optional rear window de-fogger. It costs less than $20.00 and the near horizontal rear window fogs quickly at the slightest hint of damp weather.

Forward visibility for the driver is quite good, and using the mirrors the rearward look is reasonable. However, the high headrests on the seats make it difficult to back out of a driveway without opening the door taxicab style. Still, once installed in the bucket seats, most drivers found the car fit just fine, and it was a problem prying the majority of the troops out of the Opel and back into their own equipment. We would give the car a high rating on comfort and the items that are inconvenient are very few.

Dollars and cents

Now for the show stopper — the price. The car as tested is standard and retails for $3495.00. The only extras available are the aforementioned rear window defroster, a fancier radio, and the automatic transmission which runs another ninety bucks. This is a most reasonable price for the GT and it really falls into the category of, "How do they do it for the money?" There are a lot of vehicles for

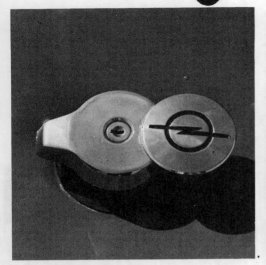

The Opel emblem swings aside to reveal the locking gas cap which also swings sideways before it will open.

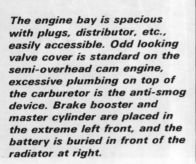

The engine bay is spacious with plugs, distributor, etc., easily accessible. Odd looking valve cover is standard on the semi-overhead cam engine, excessive plumbing on top of the carburetor is the anti-smog device. Brake booster and master cylinder are placed in the extreme left front, and the battery is buried in front of the radiator at right.

sale in the $3500 price range, but there are few with the sleek looks, dandy performance, and sporting instincts of the Opel GT at that price tag or for much more money. Yet in spite of its zingy looks, maintenance costs should be low on this car. The 1.9 liter engine has been going for some time in other models, and the majority of components are borrowed from various members of the Opel family. Fuel economy is well above average, and the car runs cool and uncomplainingly in the toughest of hot weather traffic. In short you could use it for commuting as well as cross country running and back road driving for fun. This bargain will no doubt be hard to buy for a few months, if the initial orders are any criteria. First-year production is slated for 10,000 units.

Summary

With the off again on again method by which the Opel GT was introduced in this country, it might have lost some of the excitement in the passing months. But quite the contrary, we liked the car when we saw the first pictures, we liked it more when we saw a pilot model on display last fall, and we really have taken to it since it has been in the driveway waiting for us to sail off into the hills for an hour or so of driving for the sheer fun of it. The GT is the kind of car you would like to keep, and, in fact several of our colleagues have been inquiring about buying the test car from Buick. As with any modern car there are a few cost type compromises that don't fit well into the general design. But the overall performance, the super slick handling, the quick and sure brakes, the exceptional (for a small car) comfort in the cockpit, and the delightful aura of the GT make it a most desirable machine. The Opel GT will make an enthusiast out of the most hidebound upholstered sled devotee if he will take the time to get behind the wheel of this latest goodie from Germany.

There is plenty of power from the engine for any situation, and there is more power available with some tinkering. Watch that 'tuning' though, since it is not legal in many states. We would be tempted to put some driving lights on the GT, change the safety harness for a better fit, and put in some push button window winders for convenience. Beyond that the engineers have everything pretty well thought out. Inevitably the car will be compared to GM's big beastie, the Corvette, in fact it may be dubbed 'the compact Corvette.' But this sleek new car from Adam Opel, AG, will not suffer from the comparison. The GT can stand firmly on its radial clad wheels and create a performance image for Opel that has been dormant for many years. ♠

Opel GT
Data in Brief

DIMENSIONS
Overall length (in.) 161.9
Overall width (in.) 62.2
Overall height (in.) 48.2
Wheelbase (in.) 95.1
Track (front) (in.) 49.4
Track (rear) (in.) 50.6
Fuel tank capacity (gal.) 13.0
Turning circle (ft.) 33.0

ENGINE
Type . . . . 4 cyl., in-line, ohc
Displacement . . . . 115.8 ci (1897 cc)
Horsepower (at 5200 rpm) 102
Torque (at 3600 rpm) 121

WEIGHT, TIRES, BRAKES
Weight (as tested lb.) 2090
Tires . . . . 165 HR 13 UniRoyal
Brakes (front) . . . . Disc 9.4 in.
Brakes (rear) . . . . Drum 9.1 in.

SUSPENSION
Front . . . . Independent, double A-arms, transverse leaf springs, tube shocks
Rear . . . . Live axle, coil springs, tube shocks

PERFORMANCE
Standing ¼ mile (sec.) 17.82
Speed at end of ¼ mile 76.46
Braking from 60 mph (ft.) 142.0

ROAD TEST

Read all about it.

BUICK MOTOR DIVISION

One thing before you immerse yourself in detail about Buick's Opel GT. Reading about it may be fun. But there's nothing like driving it. We invite you to stop by any Buick-Opel showroom. This year, we intend to light your fire. Opel GT.

Dimensions
Wheelbase.................95.7 in.
Overall height............48.2 in.
Overall width.............62.2 in.
Overall length...........161.9 in.
Overhang front...........36.5 in.
Overhang rear............29.7 in.
Front seat head room.....35.0 in.
Front Track..............49.4 in.
Rear Track...............50.3 in.
Turning diameter.........33.0 ft.
Curbweight
   with 1.1 SR engine......1881 lbs.
   with 1.9 engine.........2109 lbs.

Engines
Standard: 1.1 SR (1100 c.c.) Four cylinder in-line O.H.V. water-cooled. Displacement 65.8 cubic inches. Compression ratio 9.2. Horsepower, maximum output, 67 at 6000 rpm. Torque, maximum output, 62 foot pounds at 4600 to 5400 rpm. Bore 2.95 inches. Stroke 2.40 inches. Exhaust emission controls. (O.E.C.S.)

Available: 1.9S (1900 c.c.) Four-cylinder in-line O.H.V. cam-in-head water-cooled. Displacement 115.8 cubic inches. Compression ratio 9.0. Horsepower, maximum output, 102 at 5400 rpm. Torque, 115 foot pounds at 3000 rpm. Bore 3.66 inches. Stroke 2.75 inches. Exhaust emission controls. (O.E.C.S.)

Standard Transmission
Four-speed manual shift with short-throw gearshift. All forward gears synchronized. Console mounted.

Automatic Transmission
Console-mounted shift lever. Available with 1.9S engine only.

Fuel Systems
1.1 SR engine-Two single barrel Solex down-drafts with manual choke. 1.9S engine-Single Solex 2-barrel down-draft with automatic choke. 14.5 gallon fuel tank capacity. Mechanical fuel pump.

Rear Axle Ratios
1.1 SR engine 4.11, 1.9S engine 3.44.

Wheels and Tires
5J x 13 steel wheels with air slots. 165HR x 13 radial ply tires are standard.

Brakes
Four-wheel brakes. Front: power disc. Rear: drum. Total effective area 49 sq. inches with 1.1 SR engine and 79 sq. inches with 1.9S engine. Mechanical parking brake acts on rear wheels.

Instruments
0.1 mile odometer. Tach. Brake warning light. Parking brake control light, gas gauge and speedometer. Oil pressure warning light and gauge. Amp indicator light and gauge.

Electrical
12 volt battery 28 amp alternator. Sealed beam headlamps. Turn signals. Backup-light. Two-speed windshield wipers. Two-speed blower. Dome light. Cigar lighter. Electric clock. AM radio.

Interior
Carpeting front and rear. All vinyl. Full bucket seats with adjustable seat backs. Head restraints, shoulder belts and seat belts. Safety door locks. Inside day-night rear view mirror.

Colors
Exterior/Interior Trim:
Strato Blue/Buckskin
Brilliant White/Red
Flame Red/Black
GT Chartreuse/Black
Sunburst Yellow/Black
Rallye Orange/Black

MARK OF EXCELLENCE

Best Import
OPEL GT

Just the decision to build the Opel GT and import it ought to win some sort of prize for the makers. With that decision, the factory cleared at least two hurdles that have sent several other makers crashing to the ground.

The Opel GT began as a show car, a styling exercise with sedan parts. Year after year, beautiful creations are rolled onto the stage, and the builders make cryptic announcements about producing the car, with an engine even, if things work out. They don't, and the car is never heard of again.

And those sedan parts. The Opel Kadett is a nice little sedan, but the noise level would give Acid Rock a bad name. Put an economy engine inside a fancy suit, and you get a pretty car with all the zip of an economy sedan, which is to say not much.

So here's the Opel GT, a show car that made it into production. Most of the lines were changed, but the appeal wasn't. During the model year we drove some unusual cars, but none attracted as much favorable attention as the GT. There's a family resemblance to Corvette in front, the obligatory ducktail spoiler in back, joined by a sleek midsection that makes the Opel GT look like the exotic import it isn't.

The sedan components are a success. The GT and its center of gravity are lower than the Kadett, which brings body roll down to a reasonable degree, and improves cornering power. The engine has the same amount of weight to pull (GT and Kadett weigh almost exactly the same), but the slippery shape allows a taller final drive gear. Top speed is up, revs per mile are down and there's less noise from wind and engine. What howl remains didn't bother us. (A decibel meter can't distinguish between the sporty snarl of a GT car and the infuriating shriek of an economy car. But we can.)

The old bit about small on the outside, big on the inside works for the GT. Only two seats, but they're good seats. No trunk lid, but there's a nice

THE BEST CARS OF 1969

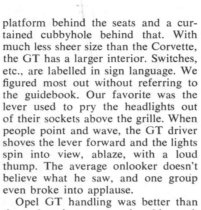
PHOTO BY DAVID GOOLEY

platform behind the seats and a curtained cubbyhole behind that. With much less sheer size than the Corvette, the GT has a larger interior. Switches, etc., are labelled in sign language. We figured most out without referring to the guidebook. Our favorite was the lever used to pry the headlights out of their sockets above the grille. When people point and wave, the GT driver shoves the lever forward and the lights spin into view, ablaze, with a loud thump. The average onlooker doesn't believe what he saw, and one group even broke into applause.

Opel GT handling was better than the sedan, but a racer has his work cut out for him. Even with the engine moved back 12 in. from its Kadett location, the GT front is heavy, and the car understeers. The inside rear wheel lifts under power, and spins in the air. In traffic or ordinary driving, the GT nips smartly about, but it's still more sedan than sports car.

So a show car made it into production without losing its looks, and sedan components can also serve as the basis for a sports car. So it's good. But why is it the best import?

What's best depends on what the import buyer wants. Low cost was the prime consideration for years, but it no longer seems to be. The imports with extras, and higher prices, are gaining ground on the cheapies all the time, and several of the major overseas builders are moving into new territory with their own versions of medium-price cars.

Speed and power? No way. The import will lose to the home-grown performance car on a dollar basis. At the very top of the scale, there are a few imported performance cars that can stay with the brutish big V-8s, but there's no such thing as an inexpensive V-12.

The imports are fun to drive, or they should be. (We can think of several exceptions.) And they look immeasurably better than the stripped sedan Detroit throws into the price wars.

The Opel GT does both, with its sporting exhaust, precise gearshift and steering, and willing engine. It darts through traffic and hums along the freeway, and the sharp good looks will be winning it glances for years. ■

SPECIFICATIONS

Wheelbase, in.	95
Overall length in.	162
width	62
height	48
No. of passengers	2
Price, basic	$3395
as tested	$3515
Frame type: Unitized.	
Front suspension: Short and long arms transverse.	
Rear suspension: Live axle, coil springs, torque control arms, track bar.	
Steering: Rack and pinion.	
overall ratio	17.4:1
turns, lock to lock	3
turning circle, ft. curb-curb	33
Curb weight	2070
Tires: UniRoyal radials 165HR-13.	
Brakes: Power assisted disc/drum.	
dia. x width, F/R	9.4/9.0 x 2.2
total swept area, sq. in.	277
Engine: IL-4	
Bore x stroke, in.	3.66 x 2.75
Displacement, cu. in.	115.8
Compression ratio	9.0:1
Rated bhp @ rpm	102 @ 3100
Clutch: Single dry plate disc.	
diameter, in.	8
Transmission: Four-speed manual, fully synchronized.	
Gear ratio 4th (1.00:1)	3.44:1
3rd (1.37:1)	4.71:1
2nd (2.16:1)	7.34:1
1st (3.43:1)	11.8:1
Lb./bhp (test weight)	23.4
Mph/1000 rpm (high gear)	19.32
Engine revs/mile (60 mph)	3100
Piston travel, ft./mile	1420
CAR LIFE Wear Index	44

ROAD TEST RESULTS

Speedometer reading @ 30 mph	28
Speedometer reading @ 60 mph	59
Top Speed (5800)	111
Acceleration 0-30 mph, sec.	3.3
0-40 mph	5.1
0-50 mph	7.4
0-60 mph	10.2
0-70 mph	13.7
0-80 mph	18.0
Standing ¼-mile, sec.	17.4
Speed at end, mph	79.0
Passing, 30-70 mph, sec.	10.4
Braking: Maximum deceleration rate ft./sec./sec.	31
No. of stops from 80 mph (at 60-sec. intervals) before 20% loss in deceleration rate	8—13% loss
Control loss: None.	
performance	excellent
Fuel consumption under test conditions, mpg	17.6
Normal cond., mpg	22-26

SEPTEMBER 1969

CAR and DRIVER ROAD TEST

Opel GT 1.9

The styling package alone puts it light years ahead of most of its competitors

There is no reason to believe that anything is sacred to those who plot the future in the war room at General Motors. Their computers do not have memory banks filled with poignant concern for Abingdon-on-Thames and Coventry or even Malvern Link. War is war. More to the point and probably more devastating, business is business and the devil take the hindmost. There is nothing personal about it. It's all done by those computers. It would be interesting to know just when it happened, but there came a day in the not too distant past when the computer, digesting the daily battle plan, decided that GM could make a comfortable profit on a $3500 sports car. Computers aren't malicious but they are cunning and the one at GM knows exactly what it takes to sell cars and make a profit. "To be a success," said the computer, *"you put the effort where the customer can see it."* In such coldly logical fashion was born the Opel GT and with that little bit of business logic behind it, it's going to steamroller Triumph and MG, maybe even Datsun and Fiat, without even feeling the lump or any remorse.

The Opel GT has two parts—the exciting, delicious, sporty styling part that you see, and the workaday mechanical part that you don't. There is precious little for mechanical sophisticates to love in the Opel's innards, but then the same accusation can justifiably be leveled at Triumphs and MGs. It's the visual Opel GT that dazzles the beholder in the showrooms and that will be enough. You, us and GM's computer all know that—with the exception of the Fiat 124—there is very little playing beneath the skin of any under-$3500 sports car, so if you buy one that at least *looks* good chances are you will come out ahead of the game. The only factor that could swing this equation around to a negative balance would be bad mechanicals. There *is* some pretty offensive hardware in the Opel but it's not that bad. It just damps our enthusiasm a little.

If you strip away the sleek bodywork of the GT you find nothing more exciting than an Opel Kadett—the same 1.9-liter engine, the same transmission, the same suspension and the same brakes. Nothing there to make sports car lovers glow with desire. Moreover, it's a sad commentary on the state of the sports car art when parts of that heritage provide performance that is more or less competitive with any sports car in the class.

It's no secret that the major inspiration for our disenchantment with Opels, both past and present, is the engine. The standard engine for the GT is a 1.1-liter Four which is too small, in a car of this appearance, to even be taken seriously. The test car was propelled by the optional 1.9-liter in-line Four, a short stroke, overhead valve design with a 9.0-to-one compression ratio and a single 2-bbl. Solex. Upon being transplanted from the Kadett it has acquired an exotically shaped, ribbed aluminum valve cover so that it at least has a sporting appearance. Purely as a source of power the bigger Opel can't be faulted. It provides a torque curve with no noticeable flat spots and is even economical (by sports car standards) in the process.

As a source of pleasure for enthusiastic drivers, however, it rates a fat zero. After living with exhaust emission standards since the beginning of 1968 you would think General Motors' money and know-how would have liberated the poor Opel from its driveability problems, but such is not the case. It still suffers from the same ailments we discovered in the last Opel Kadett we tested (February, 1968). When you lift your foot off the accelerator it seemingly takes forever for the revs to drop which means no engine braking—something like a freewheeling Saab, only you can't lock it out when you're tired of it. Moreover, it can never decide where it wants to idle. Sometimes 1500 rpm is as low as it will go and, as often as not, it reconsiders at that point and drops down to 1000 rpm. All the while it's trying to make up its mind, it's broadcasting the traditional Opel valve train clatter that would be considered rude even from a Farm-All. This is just not the stuff sports cars are made of, particularly German ones. It's almost embarrassing to wind out through the gears for fear someone will hear the ruckus coming from under the aerodynamic nose and expose you and your toy as some kind of role-playing fantasizers. Let's just say that in a car with an appearance as exciting as that of the Opel GT the engine presents a true dilemma. Combine the engine's unwillingness to slow down with an over eager clutch and you end up with a minor whiplash occasion on every upshift.

Even though the Opel's engine suffers from an advanced case of bad manners it does provide contemporary performance.

SEPTEMBER 1969

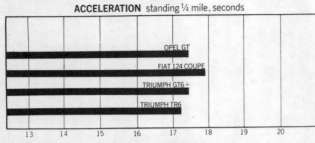

ACCELERATION standing ¼ mile, seconds
- OPEL GT
- FIAT 124 COUPE
- TRIUMPH GT6+
- TRIUMPH TR6

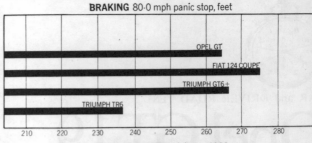

BRAKING 80-0 mph panic stop, feet
- OPEL GT
- FIAT 124 COUPE
- TRIUMPH GT6+
- TRIUMPH TR6

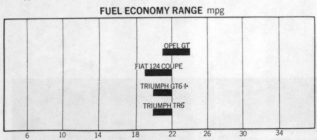

FUEL ECONOMY RANGE mpg
- OPEL GT
- FIAT 124 COUPE
- TRIUMPH GT6+
- TRIUMPH TR6

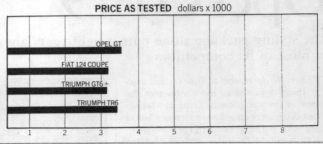

PRICE AS TESTED dollars x 1000
- OPEL GT
- FIAT 124 COUPE
- TRIUMPH GT6+
- TRIUMPH TR6

OPEL 1.9 GT

Importer: Buick Motor Division
General Motors Corporation
1051 E. Hamilton Avenue
Flint, Michigan 48550

Vehicle type: Front-engine, rear-wheel-drive, 2-passenger

Price as tested: $3524.26
(Manufacturer's suggested retail price, including all options listed below, Federal excise tax, dealer preparation and delivery charges, does not include state and local taxes, license or freight charges)

Options on test car: Base car, $3395.00; 102 hp engine, $99.26; dealer preparation, $30.00

ENGINE
Type: 4-in-line, water-cooled, cast iron block and heads, 5 main bearings
Bore x stroke..3.66 x 2.75 in, 93.0 x 69.8 mm
Displacement................115.8 cu in, 1900 cc
Compression ratio................9.0 to one
Carburetion....................1 x 2 bbl Solex
Valve gear........Chain-driven cam in head, mechanical lifters
Power (SAE)............102 bhp @ 5200 rpm
Torque (SAE)............121 lb-ft @ 3600 rpm
Specific power output........0.88 bhp/cu in, 53.8 bhp/liter
Max recommended engine speed...6100 rpm

DRIVE TRAIN
Transmission..............4-speed, all-synchro
Final drive ratio..............3.44 to one

Gear	Ratio	Mph/1000 rpm	Max. test speed
I	3.43	5.7	34 mph (6100 rpm)
II	2.16	9.1	55 mph (6100 rpm)
III	1.37	14.3	86 mph (6100 rpm)
IV	1.00	19.6	97 mph (5000 rpm)

DIMENSIONS AND CAPACITIES
Wheelbase..................................95.7 in
Track, F/R............................49.4/50.6 in
Length..................................161.9 in
Width...................................62.2 in
Height..................................48.2 in
Ground clearance.........................5.1 in
Curb weight.............................2100 lbs
Weight distribution, F/R..............54.8/45.2%
Battery capacity...............12 volts, 44 amp/hr
Alternator capacity....................300 watts
Fuel capacity..........................13.2 gal
Oil capacity............................2.9 qts
Water capacity..........................6.0 qts

SUSPENSION
F: Ind., unequal length control arms, transverse leaf spring
R: Rigid axle, trailing arms, coil springs, panhard rod

STEERING
Type........................Rack and pinion
Turns lock-to-lock.......................3.2
Turning circle curb-to-curb...........34.6 ft

BRAKES
F:................9.4-in disc, power assist
R:...........9.1 x 2.0-in drum, power assist

WHEELS AND TIRES
Wheel size....................13 x 5.0-in
Wheel type....................Stamped steel
Tire make and size......Uniroyal 165 HR 13
Tire type..............Radial ply, tube type
Test inflation pressures, F/R........24/24 psi
Tire load rating.....1080 lbs per tire @ 38 psi

PERFORMANCE
Zero to   Seconds
30 mph..............................2.9
40 mph..............................4.3
50 mph..............................7.0
60 mph.............................10.1
70 mph.............................13.5
80 mph.............................17.6
90 mph.............................24.5
Standing ¼-mile..........17.4 sec @ 79.7 mph
Top speed (estimated)..............110 mph
80–0 mph..................164 ft (0.81 G)
Fuel mileage.....21–24 mpg on premium fuel
Cruising range.....................278–317 mi

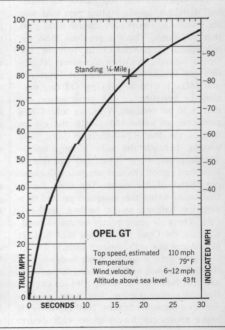

OPEL GT
Top speed, estimated   110 mph
Temperature   79°F
Wind velocity   6–12 mph
Altitude above sea level   43 ft

CAR and DRIVER

Low end torque is strong enough so that you don't have to downshift every time you drop a few hundred rpm, and the car will accelerate through the standing quarter-mile in slightly under 17.4 seconds with a terminal speed of 79.7 mph. Although the elapsed time is slightly longer than both the TR6 and the MG-C the terminal speed is about 2 mph faster, which could be construed as evidence that the Opel GT has a more favorable power-to-weight ratio than the others. Wheel spin was a definite problem during the acceleration tests so it is very likely that with a better set of tires the Opel would be marginally quicker than its British competitors.

Triumphs and MGs have nothing to worry about in the handling department, however. It's not that the Opel is dramatically inferior—again a comment on the sports car state of the art—none of the others, with the exception of the Fiat 124, is very good. The Opel is merely competent in routine sports car maneuvers and untidy if you really push it. There is nothing alarming about its suspension. The front is unique in that it uses a conventional unequal-length control arm configuration with no anti-sway bar and a transverse leaf spring instead of coil springs. Trailing arms and a Panhard rod are used on the rigid rear axle, along with coil springs. The first hint of a handling problem comes with a check of weight distribution. Of the Opel's 2100-pound curb weight, almost 55% is perched on the front wheels. This is by no means insurmountable but evidently very little effort was expended on suspension tuning. The expected initial understeer melts into an unresponsive dilemma as you approach the limit—a limit which comes on like a governor as the inside rear wheel lifts to interrupt the flow of torque to the road (the Opel GT does not have a limited-slip differential)—and the tail swings wide, balanced on one wheel. Even amateur athletes, like a Buick G-5400, can drive circles around it.

In handling, the Opel is very much like the Datsun 2000 which hasn't won any accolades either. There is no doubt that a drastic improvement could be made by the simple addition of front and rear anti-sway bars of the appropriate rate. Wider wheels and stickier tires would be a welcome improvement too. The Uniroyal 165 HR 13 radials on the standard 5-inch wide Opel wheels are never up to the task, whether it be acceleration, handling or braking. Even if you only care about cosmetic improvement, bigger balloons have a lot to offer and there appears to be plenty of room in the wheel wells for tires several sizes larger.

No matter how good a car's brakes may be it is still the tires that make up the final link between the car and the road. The Opel's braking system, power assisted with discs in front and cast-iron drums in the rear, performed smoothly and without fade throughout the test, even though the 80-0 mph stopping distances were longer than what we would expect of a modern sports car. In this case, the tires seem less capable than the brakes of maintaining a high deceleration rate.

The computers said to "put the effort where the customer can see it." As a result the Opel GT has two parts—the exciting, delicious sporty styling part that you see, and the workaday mechanical part that you don't

By this time you should understand that the mechanical Opel is not a device of rare beauty, but there is one brilliant exception —the transmission. It's a fully synchronized 4-speed like almost every other one in the car world these days, but it works very well indeed. The action is smooth and crisp with very low effort and definitely lives up to the Opel's German heritage. Still, we weren't entirely satisfied with the selection of ratios. They are very much in the European tradition, with third gear being very close to fourth and an uncomfortably large gap between second and third. A long third gear is extremely useful in Europe since passing is invariably a hazardous situation and you need all the help you can get. Driving conditions are far different in this country, however, and the gulf between second and third gets annoying in traffic.

If we found only a few bright spots in the mechanical Opel GT the opposite is more nearly correct of the visual Opel. In this day of false eyelashes and plastic leather it's little wonder that GM's computer predicted success for a car that looked sleek and sporty whether or not it was. Once inside the GT's cockpit you could easily believe you were in a Lamborghini or a Corvette or even an Apollo capsule— certainly nothing less than the latest in grand touring cars.

The bucket seats have an adjustable backrest angle and the small diameter wood-rimmed steering wheel is just the proper arm's length away. The instruments are well positioned—two large round dials for the speedometer and tachometer directly in front of the driver and the smaller gauges aimed at him from the center of the instrument panel.

Head room, leg room and elbow room are all more than adequate, but the wide transmission tunnel does force the driver and passenger to sit with their legs pointing slightly toward the outboard side of the car. The only serious mistake in the interior layout concerns the positioning of the pedals. Although there is plenty of room down there, the accelerator pedal is located so close to the brake that it is almost impossible to press the latter without getting both—an unforgivable situation which endowed the Opel with a truly treacherous nature in traffic. Anyone taking delivery of a GT should insist that the dealer relocate the accelerator closer to the tunnel before he even considers driving it out into the crowded world.

Aside from that one genuinely serious problem with controls, the GT is a relatively convenient sports car. The doors curve high into the roof so, despite its low 48.2-inch overall height, entry and exit present no problems. The flow-through vent system works reasonably well even though it never operates at its full capacity unless you crack open a side window. Like the Corvette, the GT substitutes a luggage compartment behind the seat for a trunk.

(Continued on page 25)

Autotest

OPEL GT 1900 (1,897 c.c.)

AT-A-GLANCE Opel's 1.9-litre coupé (1.1-litre engine available as alternative). High-speed performance impressive, and 100 mph cruising very good. Wind noise good with windows closed, otherwise noisy. Ventilation poor. Steering very precise. Strong understeer but predictable handling. Good brakes. Limited accommodation but great fun. Economical.

MANUFACTURER
Adam Opel AG, Rüsselsheim-am-Main, West Germany.

UK CONCESSIONAIRES
General Motors Ltd., P.O. Box 69, 23 Buckingham Gate, London, S.W.1.

PRICES
Basic	£1,387	0 0
Purchase Tax	£426	1 11
Total (in G.B.)	£1,813	1 11

EXTRAS (inc. P.T.)
Heated rear window, Alternator, and laminated windscreen £47 8 10
Fitted to test car

PRICE AS TESTED £1,860 10 9

PERFORMANCE SUMMARY
Mean maximum speed	115 mph
Standing start ¼-mile	18.6 sec
0-60 mph	12.0 sec
30-70 mph through gears	12.8 sec
Typical fuel consumption	30 mpg
Miles per tankful	360

FRANKFURT Show four years ago was the scene for the introduction of a strikingly sleek sports coupé prototype—an unexpected departure from the formerly staid and ultra-conventional cars to emerge from GM's subsidiary in Germany, the Opel works at Russelsheim. Originally it was shown with the Opel Rekord engine. The car was redesigned and production started in 1968, the bodies being built in France by Brissoneau and Lotz, and 1.1- or 1.9-litre four-cylinder engines are offered. In 1.9-litre form—the only version imported to the UK—the Opel GT has 90 bhp (DIN) available at 5,100 rpm, and has the kind of performance which its racy looks suggest.

As no examples of the car are available for full test over here we went to the factory to collect and test a works demonstrator in Germany. Not surprisingly it proved in its element as a motorway car, cruising easily at any speed up to 100 mph, and with an impressively high maximum of 115 mph, showing that the body shape is as aerodynamically efficient as it looks. We drove the car hard and far for three days and although it is by no means faultless we came to like it very much, always feeling that the car's behaviour lived up to its impressive appearance in every way.

Opening the bonnet reveals an unusual shape to the engine rocker cover, with its ribbed top and droop front. Beneath it is conventional inclined rocker gear for the valves, but the camshaft is mounted in the head, chain driven, and working cam followers directly against the rockers, without any long pushrods. Exhaust is on the right of the engine, and carburation (on the same side) is by a two-stage Solex unit. The engine is mounted well back near the bulkhead, and there is a lot of wasted space ahead of the radiator in the exaggerated snout. We even stowed the fifth-wheel towing bracket and small items of test gear there (without obstructing air flow to the radiator) which eased the problem of carrying our luggage and equipment in the confined space behind the seats. There is no boot—only a small trough partly filled by the spare wheel and toolkit, with access from within the car, revealed after unclipping a soft cover.

Starting is rather typical of engines with this carburettor in that the induction seems to get flooded after switching off and some full throttle scavenging on the starter is needed before it will fire again; from cold it likes the richer mixture, and goes off with a roar straight away, on the automatic choke. It is a very smooth and free-revving power unit, and although the gearing is not unduly high, it sounds happy to sustain indefinitely the 5,000 rpm engine speed which corresponds to 100 mph. At higher speeds, noise level builds up rapidly, and there was some thrash at maximum speed. The rev counter has an amber section beginning at 5,800 rpm, and red begins at 6,100 rpm. The upper limit is permissible for short spells and allows maxima of 32, 52 and 82 mph in the very well-spaced indirect gears. On the test car the rev counter was 10 per cent over-reading.

For drag starts it is easy to let the clutch in with a bang at about 5,000 rpm, when wheelspin gives a quite quick and smooth getaway; but we found the times were better if the clutch was fed in more gently. It took this treatment repeatedly without complaint, and reasonable figures obtained include a standing quarter-mile in 18.6sec, and from rest to 90 mph in 30.2sec; 100 mph is reached just over 11sec later, in 41.5sec. The power is well sustained at the higher speeds, the low wind resistance again helping to make the performance figures more impressive at the top end; from rest to 60 mph takes 12sec—the same as, for example, a Triumph Vitesse.

Good fuel economy

With this efficient engine and the GT's sleek body shape, it is not surprising that fuel economy is very good indeed. After correction for the small distance recorder error, 26.7 mpg was obtained on a long section of cruising at 90-100 mph on the *autobahn*, and including all the performance testing. In ordinary use, up to 34 mpg is easily obtained, and the 12-gallon fuel tank gives a range of at least 300 miles.

A short and positive gear change is fitted which is quick and pleasant to use in ordinary work but a bit too stiff when hurried. The gear knob is badly shaped, with sharp edges which dig into the hand when moving it quickly. Acceleration testing was carried out on a very hot afternoon, and occasionally our runs were interrupted by misfiring due to vapour lock.

From inside, the Opel GT presents as exciting an aspect as it does from outside. The driver and his passenger (definitely two's company, three's an impossible crowd in this confined interior) sit so low they almost feel they may touch the ground on the bigger bumps, and ahead of the windscreen are the sleek curves of the wheel arches. Between them the bonnet line falls sharply down to the road, visible only a few feet in front of the car. Forward view is spoilt slightly by a clumsy jumble of centre-parked overlapping wipers at the base of the screen, the offside blade having also an elaborate aerofoil which keeps it against the glass at speed. The wipers are turned on by a rocker switch, and have two speeds, or work at fast speed in conjunction with the washers by

Behind the seats is a flat platform for luggage, and at the back a pvc cover unclips and lifts up to reveal the spare wheel and space for tools

Although the seating position is very low there is quite a good view forwards; but the high seat backrests restrict rear vision rather severely

Above: There is no access to the luggage space, except rather awkwardly through the doors, with the seat backrests pushed forward. The fuel filler cap spins freely when locked. *Below:* The bonnet bulge clears the carburettor. Wipers are centre-parking, and their pivots are concealed beneath the bonnet

Above: The engine compartment is well laid out, and the battery and brake servo, plus mechanism for the elevating headlamps, are in the space ahead of the radiator. *Below:* The little Opel Coupé presents a very attractive profile and is aerodynamically efficient. Rear quarter windows do not open

OPEL GT 1900 (1,897 c.c.)

AUTOCAR 11 September 1969

ACCELERATION

SPEED MPH TRUE / INDICATED	TIME IN SECS
30 / 35	3.6
40 / 44	5.7
50 / 53	8.3
60 / 62	12.0
70 / 72	16.4
80 / 82	21.8
90 / 92	30.2
100 / 103	41.5
110 / 113	58.3

SPEED RANGE, GEAR RATIOS AND TIME IN SECONDS

mph	Top (3.44)	3rd (4.7)	2nd (7.42)	1st (11.8)
10-30	—	7.8	4.7	3.2
20-40	10.2	6.9	4.2	—
30-50	10.4	6.9	4.7	—
40-60	10.2	7.1	—	—
50-70	11.2	8.1	—	—
60-80	13.2	9.7	—	—
70-90	14.4	—	—	—
80-100	18.6	—	—	—
90-110	28.1	—	—	—

Standing ¼-mile 18.6 sec 75 mph
Standing kilometre 34.0 sec 94 mph
Test distance 786 miles
Mileage recorder 2 per cent over reading

PERFORMANCE
MAXIMUM SPEEDS

Gear	mph	kph	rpm
Top (mean)	115	185	5,800
(best)	116	187	5,850
3rd	88	142	6,100
2nd	56	90	6,100
1st	35	56	6,100

BRAKES
(from 70 mph in neutral)
Pedal load in lb for 0.5g stops

1	40-30	6	50-55
2	40-35	7	55
3	45-40	8	55
4	45-50	9	55
5	50-55	10	55

RESPONSE (from 30 mph in neutral)

Load	g	Distance
20lb	0.26g	116ft
40lb	0.51g	59ft
60lb	0.71g	42ft
80lb	0.83g	36.3ft
Handbrake	0.35g	86ft
Max. Gradient	1 in 3	

CLUTCH
Pedal 35lb and 6in.

MOTORWAY CRUISING
Indicated speed at 70 mph .... 72 mph
Engine (rpm at 70 mph) .... 3,540 rpm
(mean piston speed) .... 1,620 ft/min
Fuel (mpg at 70 mph) .... 34.1 mpg
Passing (50-70 mph) .... 8.1 sec

COMPARISONS

MAXIMUM SPEED MPH
Opel GT 1900 .... (£1,813) 115
TVR Vixen S2 1600 .... (£1,487) 109
Triumph GT6 Mk 2 .... (£1,148) 107
MGB GT .... (£1,217) 101
Matra M530A 1700 .... (£2,160) 95

0-60 MPH, SEC
Triumph GT6 .... 10.0
TVR Vixen S2 .... 10.5
Opel GT 1900 .... 12.0
MGB GT 13.6
Matra M530A 15.6

STANDING ¼-MILE, SEC
TVR Vixen S2 .... 17.2
Triumph GT6 .... 17.3
Opel GT 1900 .... 18.6
MGB GT 19.1
Matra M530A 19.9

OVERALL MPG
Opel GT 1900 .... 28.7
Matra M530A 26.9
TVR Vixen S2 .... 26.5
Triumph GT6 Mk 2 .... 25.2
MGB GT 22.8

GEARING (with 165 HR13in. tyres)
Top .... 19.8 mph per 1,000 rpm
3rd .... 14.5 mph per 1,000 rpm
2nd .... 9.2 mph per 1,000 rpm
1st .... 5.8 mph per 1,000 rpm

TEST CONDITIONS
Weather: Brilliant sunshine. Wind: 5 mph. Temperature: 30 deg. C. (86 deg. F). Barometer 30.4 in. hg. Humidity: 30 per cent. Surfaces: Dry concrete and asphalt.

WEIGHT:
Kerb weight 18.8 cwt (2,107 lb—956 kg) (with oil, water and half full fuel tank.) Distribution, per cent F, 54.2; R, 45.8. Laden as tested: 22.3 cwt (2,500lb—1,130kg).

TURNING CIRCLES:
Between kerbs L, 33ft 4in.; R, 33ft 8in. Between walls L, 36ft 2in.; R, 36ft 6in. Steering wheel turns, lock to lock 3.0.

Figures taken at 1,350 miles by our own staff near Munich.

PEL GT 1900 (1,897 c.c.)

AUTOCAR 11 September 1969

CONSUMPTION

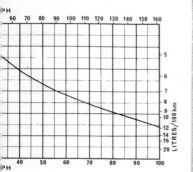

FUEL
(constant speeds—mpg)
- 0 mph 57.2
- 0 mph 48.2
- 0 mph 43.3
- 0 mph 37.8
- 0 mph 34.1
- 0 mph 30.9
- 0 mph 27.5
- 0 mph 24.2

pical mpg 30 (9.4 litres/100km)
culated (DIN) mpg 31.0 (9.1 litres/100km)
erall mpg 28.7 (9.8 litres/100km)
de of fuel Premium, 4-star (min. 97 RM)

L
nsumption (SAE 20) . . . . negligible

SPECIFICATION
FRONT ENGINE, REAR-WHEEL DRIVE

ENGINE
- Cylinders . . . 4, in line
- Main bearings . 5
- Cooling system . Sealed coolant, pump, fan and thermostat
- Bore 93mm (3.66in.)
- Stroke . . . 69.8mm (2.75in.)
- Displacement . 1,897c.c. (115.8 cu.in.)
- Valve gear . . Overhead valves and rockers, camshaft in head
- Compression ratio 9.5-to-1 Min. octane rating: 98RM
- Carburettor . . One Solex 2.choke 32DIDTA-4
- Fuel pump . . AC mechanical
- Oil filter . . Full flow, renewable element
- Max. power . 90bhp (net) at 5,100 rpm
- Max. torque . 108 lb.ft (net) at 2,800 rpm

TRANSMISSION
- Clutch . . . Fichtel and Sachs diaphragm spring, 8in. dia
- Gearbox . . . 4-speed all-synchromesh
- Gear ratios . . Top 1.0
  - Third 1.366
  - Second 2.156
  - First 3.43
  - Reverse 3.32
- Final drive . . Hypoid bevel 3.44 to 1

CHASSIS and BODY
- Construction . Integral with steel body

SUSPENSION
- Front . . . Independent with wishbones and transverse leaf spring. Telescopic dampers
- Rear . . . Live axle with trailing arms and coil springs; Panhard rod; gas-filled telescopic dampers

STEERING
- Type . . . Rack and pinion
- Wheel dia. . 13.8in.

BRAKES
- Make and type ATE discs front, drums rear (dual circuit)
- Servo . . . ATE vacuum, standard
- Dimensions . F 9.4in. dia. discs; R 9in. dia 1.97in. wide shoes
- Swept area . F 212 sq. in.; R 112 sq. in. Total 324 sq. in. (344 sq. in./ton laden)

WHEELS
- Type . . . Pressed steel disc, 4-stud fixing, 5in. wide rim
- Tyres—make . Continental
  - —type . Radial ply tubed
  - —size . 165 HR 13in.

EQUIPMENT
- Battery . . . 12 volt 44 Ah
- Alternator . . 28-amp a.c.
- Headlamps . . 45/40-watt (total) plus 55-watt Halogen pencil lamps on main beam
- Reversing lamp . Standard
- Electric fuses . 7
- Screen wipers . 2-speed, self parking
- Screen washer . Standard, electric, linked to wipers
- Interior heater . Standard, fresh air, water valve
- Heated rear window . . Extra
- Safety belts . Standard
- Interior trim . Perforated PVC seats, PVC headlining
- Floor covering . Carpet
- Jack . . . Screw pillar
- Jacking points . 4, under sills
- Windscreen . Laminated

MAINTENANCE
- Fuel tank . . 12.1 Imp. gallons (no reserve) (55 litres)
- Cooling system . 10.6 pints (including heater)
- Engine sump . 4.8 pints (2.75 litres) SAE 20. Change oil every 3,000 miles. Change filter element every 6,000 miles.
- Gearbox . . . 1.9 pints SAE 80EP. No change necessary
- Final drive . . 1.9 pints SAE 90 Hypoid. No change needed after 600 miles.
- Grease . . . None required
- Tyre pressures . F. 26; R. 26 psi (all conditions)
- Max. payload . 485lb (220kg)

PERFORMANCE DATA
- Top gear mph per 1,000 rpm 19.8
- Mean piston speed at max. power . . . 2,340
- Bhp per ton laden 95.7

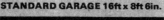
STANDARD GARAGE 16ft x 8ft 6in.

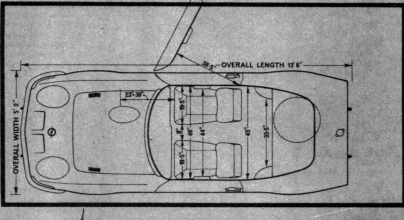

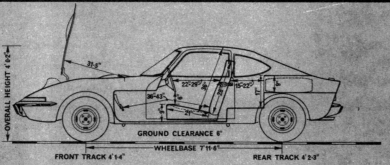

CALE 0.3in. to 1ft
ushions uncompressed

AUTOTEST OPEL GT...

pressing a button in the end of the headlamps flasher switch. It is a nuisance that the wipers cannot be lifted clear for cleaning the screen, as their arms foul the end of the bonnet.

Instruments are deeply recessed, though with flat glasses, and the speedometer is often illegible because of reflections. Minor instruments are angled towards the driver. Also in view from the driving seat is the bonnet bulge which is not just for styling, but gives clearance for the carburettor. The steering wheel is attractively dished, has stainless steel spokes and a wooden rim. The button in the centre of the boss is pressed for the characteristically high pitched horn.

The seats have very high backrests, extending sufficiently to serve as head rests. The squabs give good lateral support and the seats generally are very comfortable on a long run. Upholstery is in perforated pvc. The rear mirror gives a good view between the seats, through the sloping rear window, which is electrically heated at extra cost. It is only when the driver turns round to look backwards for reversing that the rear vision seems terribly restricted by the backrests and sloping rear quarters. Reclining backrests are standard.

Rack and pinion steering is used, and quite a lot of reaction comes back on bumps but precision is excellent and the driver really feels in control. At speed, the car responds to tiny steering corrections, and the control remains light at manoeuvring speeds. Turning circles are cumbersome. Suspension is an unusual mixture, having wishbones and a transverse bottom-mounted three-leaf spring at the front, and live rear axle on coil springs, with trailing arms and Panhard rod. There are optional anti-roll bars for both front and rear suspensions (not fitted on the test car). The coil springs are of varying diameter to give progressive rate and compensate for load variations. The ride is always very level and taut, and the car flies over rough roads with crisp, well-damped and lively suspension movements—a successful compromise between being firm and sporting yet still not harsh—and the body is impressively rigid. There are no rattles, and the suspension also works very quietly. Road noise is well insulated.

On corners there is remarkably little roll, and it takes a lot of power in a low gear to counter the natural understeer. The car always feels as though it wants to go straight on if a corner is taken too hard, but the tail will come round in easily controllable fashion if the driver provokes it with enough power and steering. On a poor surface the back jumps outwards with rear wheel hop. The Opel proved very secure and manageable on the many long, fast curves on the *autobahnen*. The quite strong understeer gives good directional stability.

Disc front brakes and large rear drums, with servo assistance, are standard, and there is a very efficient between-seats pull-up handbrake. The brakes are very progressive, and there is always reassuringly good efficiency available for moderate pedal loads. Maximum efficiency was a disappointing 83 per cent g with rear wheels locked and front ones just starting to skid, but partly this may be because we could not match abroad the very good test surfaces normally used for brake measurements at MIRA.

Scorching weather throughout the short time of the test emphasized the weakest aspect of the Opel GT—its very poor ventilation. An attempt is made at providing extraction above the rear window, but it is almost totally ineffective, and unless a window is open scarcely any air comes through the facia vents or the footwell inlets, which have large butterfly valves easily knocked closed when getting in or out. There are no front quarter vents, and the rear quarter windows are fixed. Until better extraction can be devised these should definitely be made to open. It is particularly unfortunate as the wind noise level is extremely good at speed with all windows closed, but it is very bad with either of the side windows open even a fraction. We found 65 mph was about the highest speed at which open windows could be tolerated.

The problem of high temperatures in the car in summer is accentuated by severe heat soak on the right side where the exhaust passes very close to the floor. The fresh air heater is standard and has water valve control; there is a two-speed fan, quiet on low speed but very noisy when full on.

Ingenious headlamp operation

Provision of faired headlamps which will not spoil the aerodynamic shape is very well answered in the GT. The headlamps are pivoted along the axis of the car, and revolve for use. To bring the headlamps into action a lever beside the gearlever is pushed firmly forward, and a dull thud is heard as the lamp units complete their turn and lock in the raised position, when they light up automatically. A small white telltale on the panel lights up if they have not fully engaged, up or down. They give good range, and adequate spread on dipped beam.

Beneath the bumper are two small pencil beam lamps, and when the headlamp flasher is used by day these lamps alone light up. They are also switched on at night in addition when the headlamps are on main beam. These small lamps seemed to be noticed rather less in daytime flashing at some of the slow holiday drivers on the motorways and better effect was obtained occasionally by pushing the lever forward and bringing the main headlamps into use. Maximum speed with the lamps raised was 3 mph lower, at 112 mph.

Inside and out, the car has very good finish and is extremely well made. One door lock proved stiff, and on the last occasion it was used, just before handing the car back, the key broke in the lock; but there were no other faults during the test mileage. The Opel GT is selling on the Continent in relatively large numbers, and we saw several others on the road during the test, many of them of course having the 1100 engine. Over here only the 1900 is imported, and it is expensive with tax and duty in Britain. It would be much improved by having better ventilation and more luggage space, with opening boot lid. It is fast, economical, handles and steers well, and is very eye-catching. □

The arrangement for bringing headlamps into use is ingenious; each lamp is pivoted longitudinally and revolves clockwise (as seen from the front). The under-bumper lamps are "ever-ready" for flashing by day, and come on with headlamps on main beam. They are also the side and parking lamps, using a separate filament

Here the headlamps have not completely revolved. When fully up, they light automatically if the ignition is on

Now the lamps are up, but not completely engaged, and the edges do not line up exactly with the panel; in this condition a warning tell-tale comes on

OPEL GT $3399 WEST COAST P.O.E.

General Motors styling on the German built Opel GT keeps the car recognizable as a member of the clan. Good engine power, fine brakes, dandy roadability, and genuine comfort put the Opel in the true Grand Touring category. Manufactured by Adam Opel, AG, Russelsheim, West Germany. Distributed in the United States by Buick Motor Division, General Motors Corporation, Flint, Michigan.

Opel cars have chalked up a very impressive sales record in the imported car market in this country in the last few years. Well planned merchandising through selected Buick dealers has put Opel near the top of the list for import sales. The basic sedan and station wagon are the volume items. Opel went after the performance *image* with their Rallye Coupe too. The Rallye is the standard coupe body with fancy paint, hotter engine, driving lights, a lot of instruments, radial tires, and coverless wheels. At the same time a *real* performance car was in development. Early in 1969 the Opel GT burst on the scene, properly certified and back-ordered at every dealership.

The Opel GT, made by German General Motors certainly shows its family bloodlines in styling. It not only looks similar to big brother Corvette, but car buffs have dubbed it with names like Corvette, Jr., or Compact Corvette. In actuality it is an individual design in the current vogue with a definite Kamm effect tail and the long, low, sloping nose needed to accommodate the conventional front engine/rear drive configuration in a low drag body. The body was wind tunnel tested before the design was finalized and the aerodynamic shape is not only attractive but efficient with little wind noise at speed. The clean body is devoid of the usual chrome doodads, stripes and so on, that most manufacturers tack onto their cars to promote the idea of performance. On the Opel GT even the engine hood louvers are real and functional. Decor trim consists of a small badge on the front fenders and the Opel emblem on the nose and tail. The all steel body is built by Brissonneau and Lotz in France; the slow down on body production due to the general strike in France last year was the real hang-up that caused the late introduction of the GT on these shores. The front bumper seems overly large for the car, but the rear bumperettes enhance the go-fast look and blend nicely with the chopped tail style and the really visible, large lights in the back.

The Opel is another car with disappearing headlights, but it is one of the few that looks halfway decent with the lights open. There is no vacuum or electrical assist to open the lights. Instead there is a mechanical cable actuated by a lever in the cockpit on the console. A healthy shove on the lever and the lights swing open 180°, and the light switch goes on at the same time.

The sporty GT is based on the Kadett chassis platform, but the front end is considerably changed from the bread and butter cars of the line. The engine is mounted several inches further behind the front wheels on the GT, and the steel body and chassis comes off with a very strong and rigid feeling. Typical of Teutonic cars, the doors are solid and close firmly and easily. The doors are cut slightly into the roof to aid the somewhat involved art of entry into any car that is just four feet high. The overall quality control is at a high level and the detail work would do credit to many larger and more expensive cars.

The heart of any sporting machine is the engine. Any car with the racy looks of the Opel GT should have performance to match. Although the car is listed as standard with the 1.1 liter engine, we have yet to see one so equipped in this country. The majority of cars destined for the export market here are fitted with the optional 1.9 liter or 115.8 cu. in. unit. It is a mere $100 extra at that, and the extra power is certainly worth it. The in-line, four cylinder is water cooled and lies somewhere between a true overhead cam layout and the more common over-

JANUARY 1970

The Opel has basic understeering tendencies and the body roll is evident in hard cornering. Roadability is quite good and the car is extremely stable in all conditions.

Instrument panel is well done with the big tach and speedo right in the line of sight. Auxiliary gauges and numerous buttons for other controls fill the center console. Lever to the left of the ashtray is manual release for the headlights.

head valve style for European four-bangers. The camshaft is actually in the head over the wedge shaped combustion chamber, but under the rocker arms. Drive is by a hydraulically tensioned duplex roller chain. The valves operate without pushrods, but hollow tappets are interposed between the cam and the rocker arms. This somewhat unusual valve gear arrangement keeps the engine height lower than average. Induction is by a single, two-throat Solex carburetor with an automatic choke. With a bore of 3.66 in. and a stroke of 2.75 in. the engine puts out 102 comparatively smog-free horsepower at 5200 rpm. The torque rating is 121 lb./ft. at 3600 rpm and the compression ratio is 9.0 to 1.

The major fault with this engine is the high noise level that has always been characteristic, even at mid-range on the rpm scale. In the GT the insulation is good and the engine is more quiet than when placed in the Rallye Coupe; but the fan racket and general roaring and thrashing is still quite noticeable at normal cruising speeds. Conversely the exhaust system of twin mufflers and dual pipes is quiet and doesn't make the expected sporting sounds. A glass pack would fix that right up. The engine bay is roomy and access to the working parts is quite good.

A four speed, all synchro gearbox is standard and there is an optional three speed automatic transmission available. The gear lever is well positioned on the center console. The smooth working, typical German synchros are hard to beat, and the linkage is quite good on the four speed. There is a lock-out for reverse gear that works off a lift-up collar near the top of the gear lever, and, for the uninitiated, the shift pattern is marked on the knob.

The disc front, drum rear brake set-up is common on European cars. The GT has excellent brakes with the 9.4 inch discs on the front, balanced well with the 9.1 inch drums at the rear. Of course it shouldn't be much of a problem to put good brakes on a one ton car. Still the Opel brakes are really fine with an ATE power boost system that is barely detectable to the driver. The pedal feels firm and progressive as it should be on a sports car.

Inside the cockpit the Opel is fairly spacious. There is no place for a back seat and no attempt at making it a 2 plus 2. The fine bucket seats seem to fit all types of physiques, and they are handsome in appearance too. There is a recliner mechanism and a full travel track for a variety of driving positions. The seats provide good support for the back and legs for long journeys. Pedals are slightly offset, but well spaced and about the right size. The three spoke, wood-

ROAD TEST

The in-line four cylinder engine rides low in the front bay and most accessories are handy for maintenance. Space is gained by placement of the battery, right, and the brake booster/master cylinder forward of the radiator. Most of the visible plumbing concerns the anti-smog devices.

rimmed steering wheel is properly sporty, but is not adjustable. It is mounted low and with the variety of seat adjustment anyone can fit into a comfortable driving position.

There is a full panel of VDO instruments in the black face white letter style. The large, round tachometer and speedometer are mounted on either side of the steering post and are in view without conscious effort. At the top of the center dash are three smaller dials that hold the ammeter and oil pressure gauge, water temperature and gas gauge. The third is a very accurate time of day clock. Under this is the standard equipment radio and a brace of buttons with the usual Continental pictures depicting their uses as light switches, warning devices, wipers, etc. Underneath all this are the heater controls. Fresh air vents are on the lower dash on each side. Air is exhausted through vents above the rear window and ventilation is quite good. The rear window has an optional defroster, a handy thing in poor weather. One dissonant note in the otherwise luxurious interior is the cheap wood grained contact paper that covers the console. There is no trunk at all in the GT, but the full space under the back window can be used for odd bits of baggage and there is an open cubby hole in front of the passenger on the dash.

The forte of the small engined sports car is fine handling and genuine roadability. Most are not famous for blistering acceleration, but the total car is usually popular because of the nimble behavior that makes full use of the engine power possible. The Opel GT is quite a fun car to drive. Despite the sedan type underpinnings, the engineers have put together a fine machine that goes well anywhere. Handling tends to initial understeer, but as cornering speeds increase the Opel becomes more neutral and sticks very well. In a tight corner the inside rear wheel will lift, but one day such options as a limited slip differential and proper sway bars will come along to make for better chassis performance. On twisty roads the GT lives up to its name, and even the ham-fisted will find it an easy car to fling around with aplomb. The steering is quick and precise from the rack and pinion unit, and the gear spacing is really fine for enthusiastic type driving. In fact the GT is quite comfortable to work with in the mountains or the flatlands.

The Opel GT is quite a departure for General Motors German branch. They haven't built a sports car at Opel since before the first World War. Despite the GT's obvious relation to the parent company in styling, it has many individual traits and finer quality control than one finds on similar priced domestic vehicles. The GT is in the $3500 price bracket with a few options such as the automatic transmission available if desired. It is quite a car for the money in its class. Overall performance is good, as are handling, power and brakes. It has comfort, and also the delightful aura that makes a GT so desirable. The GT is creating a performance image for Opel that has been lacking for many, many years. ♠

Side by side the Opel displays a styling heritage from big brother Corvette. Fender lines are quite similar and overall styling appears directly related.

Opel GT

Data in Brief

DIMENSIONS
Overall length (in.) 161.9
Wheelbase (in.) 95.1
Height (in.) 48.2
Width (in.) 62.2
Tread (front in.) 49.4
Tread (rear in.) 50.6
Fuel tank capacity (gal.) 13
Luggage capacity (cu. ft.) 6.6
Turning diameter (ft.) 33.0

ENGINE
Type . . . . OHV in-line, 4
Displacement (cu. in.) 115.8
Horsepower (at 5200 rpm) 102
Torque (lb./ft. at 3600 rpm) 121

WEIGHT, TIRES, BRAKES
Weight (curb lb.) 2090
Tires . . . . radial 165 HR 13
Brakes, front . . . . disc
Brakes, rear . . . . drum

SUSPENSION
Front . . . . independent, double A arms,
transverse leaf spring, tubular shocks
Rear . . . . . . live axle, coil springs, tubular shocks

THE OPEL GT

A sports 2-seater

IT WAS at the Earls Court Motor Show in 1968 that I first saw the Opel GT and it looked so nice and felt so right when you sat in it, that I was eager to try it on the road. It really is amazing to think that nearly two years have gone by before the opportunity to drive one arrived, but I suppose this is mainly because the world of Opel and their family cars is one that I do not frequent. However, over the past eighteen months I have been conscious of all sorts of noises coming from Adam Opel AG in Germany and motor-racing colleagues have frequently disappeared off to the Opel factory muttering words like "significant, improved, very good", to which I have usually replied "Don't tell me Opel are beginning to make good cars, they have never made a car to interest me, I wonder why they are starting now?" Further noises on the grape-vine kept saying that General Motors of America were getting a bit sick of Ford receiving so much publicity from competitions and that Ford were using racing to build a successful sales image. This obviously meant that General Motors, in one form or another, were going to join the motor racing band-wagon and develop the sporting image. That Opel have produced a sleek, purposeful-looking little sports coupé, called the GT, can hardly be coincidence. As soon as I saw the Opel GT I felt it must sell, no matter how good or bad it was, especially in Europe where sales and service for GM products is widespread. All over Europe the GT is appearing in larger and larger numbers, being available in 1,100 c.c. and 1,900 c.c. form, and while the former must be a bit gutless the latter can't be bad, especially as the 1,900 c.c. model is quite a sophisticated single o.h.c. layout, with an over-square engine of 93 × 69.8 mm. bore and stroke.

It was fortunate that just as I was having a weekend off from European racing, with the idea of taking a look at two other sports, namely a round in the Shell/RAC Hill-Climb Championship and an Autocross, the Editor found himself with more cars than he knew what to do with so he offered me the Opel GT, and I did not hesitate. After using the little silver coupé, with left-hand drive and automatic gearbox for a few days I was very reluctant to give it back and I now realise why those various colleagues had returned from Adam Opel muttering "hmm, significant, remarkable, very good". Until now the name Opel has brought a tolerant smile to the lips of motoring enthusiasts and they have gone on talking about Lotus Elans, MG-Bs, Alfa Romeos, even TVRs and Gilberns, but the Opel GT has changed all this. During the weekend the most convincing thing was when I lent it to a Lotus enthusiast, who knows what he wants in a sports car, to the point of building his own, and who only knows one way to drive. He came back with a very puzzled and quizzical look on his face, and when I asked him what he thought was wrong with it, he said quietly and still rather puzzled "There's nothing wrong with it". Of course, we were talking about the conception of the car, the way it performed, the way it steered and went round corners, how it braked or changed directions and so on; in other words its roadability. On "nit-picking" details there are lots of things wrong, like the window winders that are practically under the seats if you have them forward on their runners, or the fact that the steering wheel spokes are rather thick and mask the rev.-counter and speedo, and that the spare wheel is inside the tail and has to be pulled out from inside the car (imagine having to put a punctured tyre and wheel back in the tail in the dark and rain, just after having run through some cow pats), or that the headlamps do not always retract completely. Looking at the Opel GT as a sports car, for the way it goes and the enjoyment you get from driving it, it has reached a very high standard and I put it in the Lotus Elan category without reservations.

Looking underneath it was reasonable to find unequal-length double-wishbones at the front and telescopic shock-absorbers, but a surprise to see a transverse leaf-spring, of three wide and thin leaves. At no time while driving the Opel GT was there any need to think about the rear axle or rear suspension, so it was another surprise to see a beam axle underneath; however, it was suspended on large-diameter coil-springs and located by forward-running radius-rods, a Panhard rod, a torque tube trunnion mounting, telescopic shock-absorbers mounted at 45-degrees and an anti-roll bar in a manner known as "well tied down" and it was all very effective on corners as well as giving a Lotus-like ride and control. Cornering was assisted by 165 HR × 13 G800 Goodyear tyres on 5J rims, and the steering is by rack-and-pinion and nice and direct. This coupled with the suspension makes sudden changes of direction at 60-70 m.p.h. no trouble at all and the car can be dodged about either for fun or to avoid wayward cats, dogs, pheasants or birds, or for traffic manoeuvres. The single overhead chain-driven-camshaft 1.9-litre engine is quoted as giving 90 b.h.p. DIN at 5,100 r.p.m. or 102 h.p. SAE at 5,400 r.p.m., with a recommended maximum of 6,000 r.p.m. using a 9.5-to-1 compression ratio.

On the road the test-car, with the three-speed automatic transmission, would show 5,000 r.p.m. in "high" on any piece of straight road, and given a bit of a run it would wind up to 5,800, where the yellow segment on the tachometer begins. An *Autobahn* would allow it to creep on up to a full 6,000 r.p.m., so the 3.44-to-1 rear axle gearing would appear to be about right, especially as the four-cylinder engine was nice and smooth at peak r.p.m. The automatic transmission, with no clutch pedal, was operated by a rather nice T-handled lever on the central backbone between the seats, moving in a simple fore-and-aft plane, from "Park" fully forward to "Hold in Low" fully back, passing through "Reverse", "Neutral", "Drive" and "Hold in Middle" on the way, with a safety catch under the "T" to prevent inadvertently selecting "Reverse". The car was best driven on the "left foot braking" principle and as the automatic transmission had all the usual goodies, like "kick-down" and "hold" positions you could drive the car as with a three-speed gearbox without a clutch. However, by the time you had played all the games with the automatic, you felt you might just as well have a decent close-ratio four-speed box like an Elan, except that by all accounts the four-speed manual box offered as an alternative on the Opel GT is not "a decent close-ratio four-speed". In fully automatic, changes took place at 5,200 r.p.m. on full-throttle or else when you eased your foot on the accelerator.

In daylight the indicator stalk on the left of the steering column is used to flash the low-mounted spot-lamps, and at night the headlamps are brought into use by a lever alongside the gear-lever. You push this forward and a mechanical linkage rotates the elliptical plates on

Continued on page 79

MOTOR ROAD TEST No. 24/70   Opel GT

For sporting gents—at a price
Stylish two-seater coupe; good performance and handling; smooth automatic transmission; poor switchgear, needs a boot lid; very expensive in Britain

It is not often that we get dream cars for test; the Opel GT is one of them. It was first seen as an Opel styling exercise at the 1965 Frankfurt Show, powered by a highly modified version of the Opel 1.9-litre ohc engine said to give a maximum speed of around 125 mph. Unlike Vauxhall's similar XVR the GT dream became reality in October 1968 when Opel released a production version. Structurally, it is based on the Kadett, using a stiffened Kadett platform and essentially the same running gear. Double wishbones are controlled by a transverse leaf spring at the front, and there's a live axle at the rear, driven through a torque tube and located by a Panhard rod and a pair of trailing radius arms. The rear coil springs, however, are made of tapered wire to give a variable rate, and a rear anti-roll bar is standard on cars sold in

PRICE: £1574 plus £483 purchase tax equals £2057. With automatic transmission as tested, £2194

the UK. Opel engineers emphasize the attention paid to aerodynamics in shaping the body—which incorporates a roll-over structure—both to reduce drag and to achieve a rearward centre of pressure for straight-line stability at high speeds. In Germany it is available with either a 1.1-litre or a 1.9-litre engine, but in Britain it is supplied only with the larger unit.

Ironically the car is a failure as an *objet de style* if judged by the severest standards. We don't consider it ugly, nor do we object to its strong familial resemblance to the Sting Ray Corvette—which the designers obviously considered to be the ultimate in sports cars—but we do find its proportions unfashionably narrow. Even the children of some of our test staff commented on this. Its stowage space is also poorly planned, since the spare wheel and tool-kit can become trapped between the petrol tank at the rear and the luggage behind the seats—a rear door of the sort to be found on the Triumph GT6 or the MGB GT is badly needed.

In this country the car is only available in left-hand drive form and at £2194 with automatic transmission it is expensive; you can get a manual Jaguar E-type for very little more, and direct rivals cost around £800 less.

Nevertheless, it is a pleasant car to drive with reasonably adequate performance and an excellent automatic transmission. It also excels where least expected in having very good roadholding and exceptionally good, vice-free handling.

Performance and economy

It pays to observe the instructions in the handbook meticulously when starting from cold: leave the accelerator pedal completely alone for the first minute after an initial dab to set the automatic choke. Premature depression of the pedal or selection of drive makes the engine stall; after that it pulls readily. Though smooth enough in its way the Opel unit is neither refined nor quiet. With its clattery valvegear—a single chain-driven overhead camshaft operates in-line inclined valves through rockers—it sounds a bit like a truck engine. Although the tachometer is red-lined at 6100 rpm it is a waste of time running the engine to this speed unless absolutely necessary when overtaking or holding a gear in a corner, since maximum power is developed at 1000 rpm less and the unit runs out of breath immediately thereafter. We tried holding the engine to maximum revs during some manual-selection standing start acceleration runs, but the times were slower than with normal automatic changes at around 5000 rpm.

Subjectively the GT feels quite fast and it gets to an indicated 100 mph reasonably quickly, albeit helped by an optimistic speedometer which is 7 mph fast at that speed. It also feels very stable at high speed, perhaps as a result of Opel's aerodynamic efforts to keep the centre of pressure to the rear and maintain down-pressure on the back wheels. But their wind tunnel work seems to have been less effective in reducing drag, since its performance is not outstanding for 90 bhp, although this may be partly due to the inevitable power losses in the automatic transmission. And at nearly 19cwt. the car is heavy for its size. Maximum speed was 106 mph and the 0-60 mph acceleration time 12sec.

Our constant-speed fuel consumption figures show the GT to be quite economical up to 70 mph, which accounts for the good touring consumption of 29.5 mpg on four-star petrol. Overall consumption was a fair 25.9 mpg, but most private owners will probably do a little better than this, say 27 mpg, giving a range from the 11-gallon tank of around 280 miles.

Transmission

This test provided a useful opportunity to assess the new GM three-speed automatic transmission built in Strasbourg: hitherto European GM cars have been equipped either with Borg-Warner automatic transmissions, or with GM units shipped over from the States.

The new three-speed-plus-torque-converter transmission impressed us with its very rapid, near-instantaneous, yet very smooth upward changes and kickdown selection. Its settings were also well chosen for a sports car, automatic changes from first to second occurring at 5200 rpm, at 5000 rpm from second to third, the equivalent speeds being 42 mph and 66 mph, representing a reasonable spacing of ratios for a three-speed box. Second gear was available at any speed up to 60 mph by kicking down, giving good overtaking.

Our main criticism of this transmission is of the violence of

PERFORMANCE

Performance tests carried out by *Motor's* staff at the Motor Industry Research Association proving ground, Lindley.

Test Data: World copyright reserved; no unauthorized reproduction in whole or in part.

Conditions
Weather: Warm and dry, wind 0-5 mph
Temperature: 62-68°F.
Barometer 29.8 in. hg.
Surface: Dry concrete and tarmacadam
Fuel: Premium, 98 octane (RM), 4 Star rating

Maximum Speeds

	mph	kph
Mean lap banked circuit	106.0	170.8
Best one-way ¼-mile	108.3	175.0
Intermediate	66	106
Low	42	68

"Maximile" speed: (Timed quarter mile after 1 mile accelerating from rest)
Mean 101.0
Best 103.2

Acceleration Times

mph	sec
0-30	4.4
0-40	6.5
0-50	9.0
0-60	12.0
0-70	16.2
0-80	22.9
0-90	31.2
0-100	45.9
Standing quarter mile	18.6
Standing kilometre	34.5

Kickdown
mph	sec.
20-40	3.9
30-50	4.6
40-60	5.5
50-70	7.2
60-80	10.9
70-90	15.0
80-100	23.0

Fuel Consumption
Touring (consumption midway between 30 mph and maximum less 5% allowance for acceleration) 29.5 mpg
Overall 25.9 mpg
(= 10.9 litres/100km)
Total test distance 1030 miles

Brakes
Pedal pressure, deceleration and equivalent stopping distance from 30 mph

lb.	g.	ft.
25	0.37	81
50	0.67	45
75	0.86	35
105	1.00	30

Fade Test
20 stops at ½g deceleration at 1 min. intervals from a speed midway between 40 mph and maximum speed (= 73 mph)

	lb.
Pedal force at beginning	30
Pedal force at 10th stop	30
Pedal force at 20th stop	30

Steering
Turning circle between kerbs: ft.
Left 33
Right 32½
Turns of steering wheel from lock to lock 2.8
Steering wheel deflection for 50 ft. diameter circle 0.9 turns

Speedometer
Indicated	10	20	30	40	50	60	70	80	90	100
True	9	17	26	37	46	56	65	74	83	93

Distance recorder 2% fast

Weight
Kerb weight (unladen with fuel for approximately 50 miles) 18.5 cwt
Front/rear distribution 55.5/44.5
Weight laden as tested 22.3 cwt

Parkability
Gap needed to clear 6 ft. wide obstruction in front

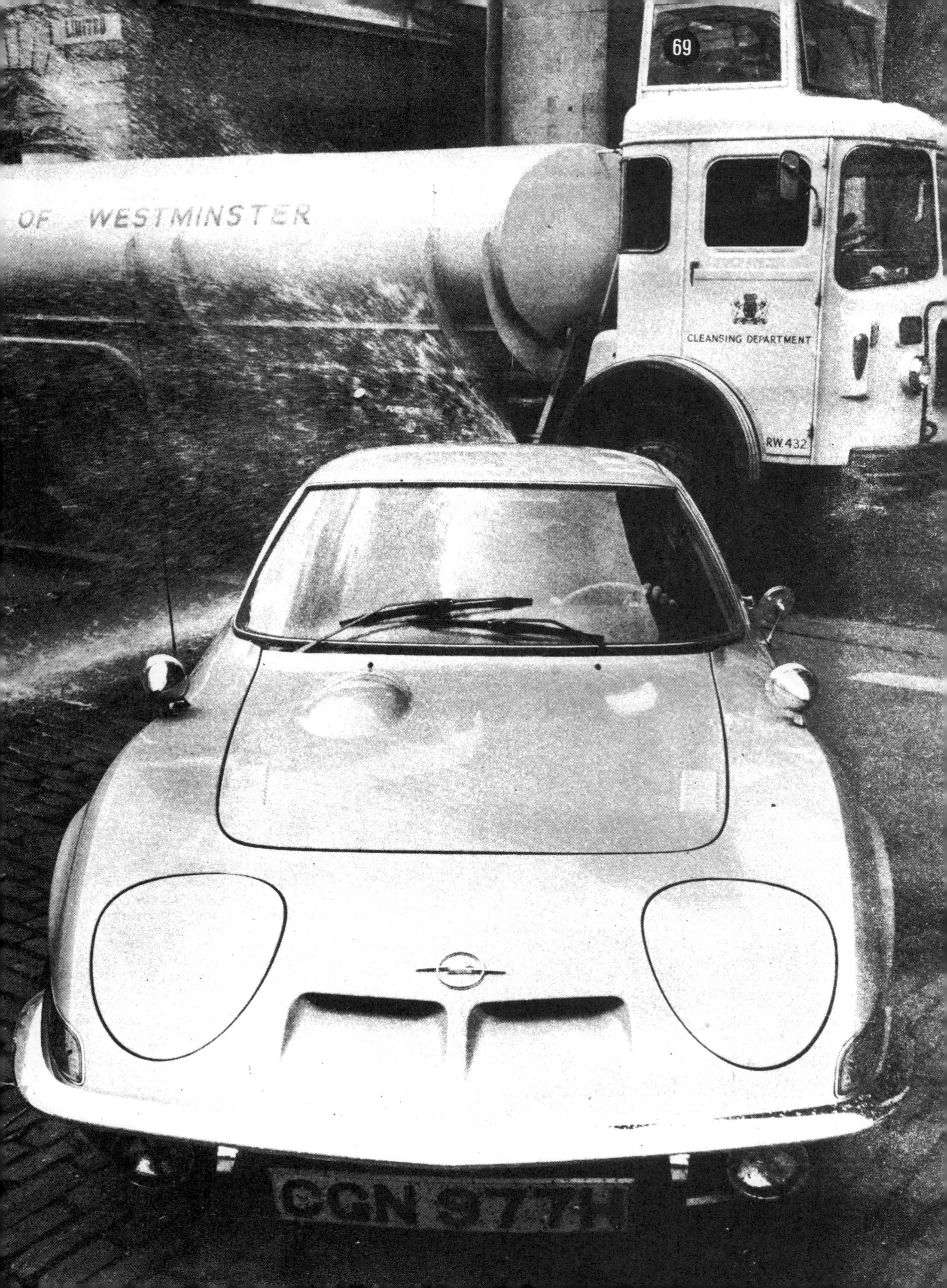
OF WESTMINSTER
CLEANSING DEPARTMENT
RW 432
CGN 977H

Above: centre of attention—but no way in to the boot. Left: left-hand drive and a typical GM facia layout. Below: with cushions for padding, there's room for two children in the back for short spells. Right: pop-eye lights swing into position when you push a lever inside

manually selected downward changes which were always accompanied by a jerk, often the shock torque was enough to squeak the back tyres. In bad conditions we felt this could unbalance the car, though it never did so in the dry. We had no quarrel with the action of the gear selector lever itself, with its lift-up T-handle. The rear axle whined quite loudly, noticeable at low speeds in traffic as on occasion was a prominent "wow-wowing" beat frequency noise. There was also some whine from the gearbox in first and second.

Handling and brakes

From a handling and roadholding point of view, the design of the Opel GT seems most unpromising. It is fundamentally a stylists' car, a sculptured shape that had to be adjusted a little to fit a Kadett platform. The double wishbone layout of the front suspension it has inherited in this way controlled by a transverse leaf spring arranged to give increasing stiffness with roll—a system that worked badly on our own Viva HA—though the torque-tube driven live rear axle is suspended on coil springs and adequately located by a pair of trailing radius arms and a Panhard rod.

But in defiance of these omens the GT has very good roadholding and exceptionally safe and vice-free handling. Although its rack and pinion steering is heavy at parking speeds, on the road it is light, accurate and high geared—2.8 turns from lock-to-lock. In conjunction with the moderate roll, the quickness of the steering cuts down wheel-twirling and makes it easy to follow sharp bends accurately and without effort. It also makes light work of the mild understeer, which is the essential handling characteristic. At the limit the car simply loses speed through tyre scrub without swinging too wide: it is almost impossible to unstick the tail in the dry. We did manage it by selecting "Low" for a sharp corner and flooring the accelerator which caused the inside rear wheel to lift and spin, but such a manoeuvre is seldom necessary or useful. The Goodyear G800 radials tended to squeal softly when mildly provoked without proportionately increasing their song during really hard cornering. Their grip was excellent.

The servo-assisted disc/drum brakes were progressive and needed moderate pressures; a push of just over 100lb. gave 1g deceleration. They were not affected by a thorough soaking in the watersplash, and they did not fade, either during our test or during fast driving on twisty roads.

MOTOR week ending June 20 1970

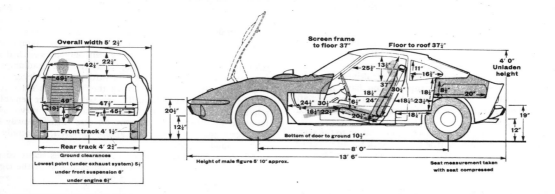

Engine
Block material	Cast iron
Head material	Cast iron
Cylinders	4 in line
Cooling system	Water
Bore and stroke	93 mm (3.66in.) 69.8 mm (2.75in.)
Cubic capacity	1897 cc (115.8 cu. in.)
Main bearings	5
Valves	sohc
Compression ratio	9.0:1
Carburetter	2 barrel 32 DIDTA-4
Fuel pump	Mechanical
Oil filter	Full flow
Max. power (net)	90 bhp at 5100 rpm
Max. torque (net)	108 lb. ft. at 2500 rpm

Transmission
Internal gearbox ratios
Top gear	1.00:1
Intermediate	1.48:1
Low	2.40:1
Reverse	1.92:1
Final drive	3.44:1

Mph at 1000 rpm:
Top gear	19.6
Intermediate	13.3
Low	8.2

Chassis and body
Construction . . . . Unitary

Brakes
Type	Twin-circuit discs/drums with servo
Dimensions	9.4in. dia front; 9.06in. dia. rear
Friction areas:	
Front	16.28 sq. in. of lining
Rear	62.77 sq. in. of lining operating on 112 sq. in. of drum

Suspension and steering
Front	Independent by unequal length wishbones and a transverse three-leaf spring.
Rear	Live axle on variable rate coil springs with two radius arms and a Panhard rod.

Shock absorbers:
Front	} Telescopic
Rear	
Steering type	Rack and pinion
Tyres	Goodyear G800
Wheels	13in.
Rim size	5J

Coachwork and equipment
Starting handle	No
Tool kit contents	Wheelbrace, jack, etc.
Jack	Screw pillar
Jacking points	One each side under sills
Battery	12 volt negative earth 44 amp.hr capacity
Number of electrical fuses	7
Indicators	Self cancelling flashers
Reversing lamp	Yes
Screen wipers	Two speed, self-parking
Screen washers	Electric
Sun visors	2
Locks:	
With ignition key	Steering
With other keys	Doors and filler cap
Interior heater	Fresh air
Upholstery	Vinyl
Floor covering	Carpet
Alternative body styles	None
Maximum load	441 lb
Maximum roof rack load	88 lb
Major extras available	Laminated screen

Maintenance
Fuel tank capacity	11 galls
Sump	5.3 pints SAE 20W50
Gearbox	3.7 pints SAE Automatic transmission fluid
Rear axle	1.9 pints SAE 90
Steering gear	Sealed for life
Coolant	10 pints (1 drain tap)
Chassis lubrication	None
Minimum service interval	3000 miles

Ignition timing	tdc
Contact breaker gap	0.016-0.020in.
Sparking plug gap	0.027-0.031in.
Sparking plug type	Bosch W200T35 or AC 43FO
Tappet clearance (hot)	Inlet 0.012in. Exhaust 0.012in.
Valve timing:	
inlet opens	44° btdc
inlet closes	86° abdc
exhaust opens	84° bbdc
exhaust closes	46° atdc
Front wheel toe-in	0.035-0.114in.
Camber angle	47' ± 1°
Castor angle	2° 40' +1° −2°
King pin inclination	6°
Tyre pressures:	
Front	26 psi
Rear	26 psi

Safety check list
Steering Assembly
Steering box position	Well back
Steering column collapsible	Yes
Steering wheel boss padded	Yes
Steering wheel dished	Yes

Instrument Panel
Projecting switches	No
Sharp cowls	No
Padding	Extensive

Windscreen and Visibility
Screen type	Toughened
Pillars padded	Yes
Standard driving mirrors	Interior and on door
Interior mirror framed	Yes
Interior mirror collapsible	Yes
Sun visors	Crushable

Seats and Harness
Attachment to floor	On slides
Do they tip forward?	Yes
Head rest attachment points	No
Safety Harness	Lap and diagonal

Doors
Projecting catches	No, recessed
Anti-burst locks	Yes
Child-proof locks	Yes

1 multi-function stalk. 2 tachometer. 3 warning light cluster. 4 mileometer and speedometer. 5 oil pressure gauge. 6 ammeter. 7 fuel gauge. 8 temperature gauge. 9 clock. 10 ignition/starter lock. 11 lights master switch. 12 dummy switch. 13 panel light switch. 14 heater temperature control. 15 dummy switch. 16 heater distribution control. 17 dummy switch. 18 hazard warning switch. 19 wiper switch. 20 backlight heater switch. 21 fan switch

Comfort and controls

For a sports car with a live axle the Opel's ride is surprisingly good, perhaps because of the variable rate rear springs. There is some low-speed joggling over small bumps, and certain short-wavelength undulations set up a restless pitching movement, changing to a gentle, well-damped float as the waves increase in length. In general though, the suspension copes well with rough surfaces at high speeds. Some of our drivers liked the firm seats which offer fair lateral support and have reclining backrests as well as an extremely good range of fore-and-aft adjustment. Others found them uncomfortable and complained of poor lumbar support.

Wheel and gear selector are well located in relation to the seat, but the handbrake is perhaps a little too far back for an effective pull in an emergency. The pedals are well placed, too, but on our test car the accelerator had some freedom to move sideways and sometimes got too close to the brake, resulting in an

occasional alarming absence of deceleration, the normal sensation of driving against the brakes being masked by the cushioning effect of the torque converter.

The layout of the minor controls is less satisfactory. There is a single stalk controlling the indicators, flash and dip, with a push-button tip which operates an effective wash-and-wipe facility. Flashing is done by the two fixed halogen lamps on each side of the grille which otherwise can only be switched on by selecting main beam and raising the retractable headlamps. This is done by firmly pushing forward a small lever in the transmission tunnel which swivels the headlamps sideways and switches them on automatically—probably a more sensible and reliable method than the electric or vacuum servo systems favoured by other manufacturers on cars with foldaway lights. The operation never fails to amuse children, especially as it makes a loud thump.

All other minor controls are rocker switches scattered across the console in two groups of three on each side of the cigarette lighter with a further smaller group below the heater controls. On our test car no fewer than three of the switches were dummy ones, demonstrating the stylists' childish desire to preserve symmetry rather than ergonomics in the layout.

With thin screen pillars and no quarterlights, forward visibility is good. The car's narrowness and upswept wings (good aiming marks) make manoeuvring easy in confined spaces—partial compensation for the bulges in the car's flanks. Visibility directly to the rear of the car is quite good except in the wet when it is badly obscured by water droplets on the near-horizontal backlight. Three-quarter rear visibility is poor.

In the hot weather of our test the interior of the car became rather warm, despite some relief obtained from the rather crude through-flow ventilation system with inlets at the ends of the facia and in the footwells. The same weather conditions prevented us from making a proper assessment of the heater, though its distribution control seemed effective and its temperature control progressive.

With all the attention Opel claim to have devoted to the aerodynamics of the GT, it's not surprising that wind noise is low. Road noise is well isolated, too, leaving as residue some whines from the transmission and the busy clatter of the engine.

Fittings and furniture

The speedometer and rev counter are large, attractive, and well located immediately in front of the driver. The auxiliary instruments, mounted at the top of the central console and angled Alfa-Romeo fashion towards the driver, include ammeter, fuel, oil pressure, and water temperature gauges, and a clock. Lower down in the console are a cigarette lighter and a hazard flasher switch, which, like the reversing lamp and an underbonnet lamp, is a standard fitting. Coathooks in the rear quarters and a large ashtray in front of the gear selector also form part of the standard equipment.

Space for oddments is limited to a small open glove compartment, which also serves as a grab handle, a small tray on the transmission tunnel, and a rear parcel shelf, large in area, but suitable only for flat objects as the steeply sloping shallow rear window is easily obscured. Behind the rear seats is a space large enough to carry two small children for short journeys or a pair of substantial suitcases. Behind this space in turn, and covered by a button-down flap, is a recess for the spare wheel and toolkit; changing a wheel may involve removing all the luggage and manoeuvring the wheel past the tip-forward backrests.

Servicing and accessibility

The Opel engine fits snugly into its compartment with some room left over in front of it. Most of the important service points such as the dipstick, oil filler cap, carburetter, distributor, alternator, brake reservoir and washer bottle are easy to get at. Servicing is required every 3000 miles at which the main job is to change the engine oil. There are some 80 Opel dealers in the UK.

The pillar jack is easy to use, but the toolkit and spare wheel are difficult to reach in a recess at the back of the luggage space behind the seats.

1 brake reservoirs. 2 radiator filler cap. 3 air cleaner. 4 distributor. 5 coil. 6 washer bottle. 7 dipstick. 8 oil filler cap

MAKE: Opel **MODEL:** 1.9 GT **CONCESSIONAIRES:** GM Ltd., Stag Lane, Kingsbury, London NW9

TUNING TEST

OPEL GT
More power from GM's German coupe

DURING LAST month we had an enjoyable break from the normal run of modified production saloons as the well-publicised, but little seen, Opel GT passed into our hands for a short period, equipped with some equally well-known Steinmetz tuning equipment. First shown as a prototype in 1965, the Opel 1.9-litre GT went into production in the latter half of 1968, and met with approval from all those who had felt that all Opels were boring, though the specification is really no more exciting than an MGB GT. However the end result is a lot more modern in style and handling than the British car, and most press reports (including our own by D.S.J. in 1970) indicated that the German GM division were onto "A Good Thing".

Klaus Steinmetz is an ex-BMW works racing team manager who left Munich several years ago to concentrate on the development of the Opel engine and chassis range, especially of the big straight-six Commodores and the four-cylinder cam-in head Kadetts of 1.9 litres (the same engine as is used in the Manta and GT, of course). In Germany and throughout the Continent, Steinmetz Opel tuning parts and knowledge have become much sought after, largely because the Kadetts have proven very successful in Swedish rallying hands, while Commodores have had their moments in international saloon car races.

The car we tried was an automatic, fitted with a straight-forward German Steinmetz conversion, whereas future converted Opels in Britain will be further modified toward smooth performance by ex-works Mini driver John Rhodes. The latter gentleman was to have driven an Opel in European Touring Car Championship events this year, but because the drive did not materialise he set up John Rhodes Tuning Ltd., Steinmetz UK distributors at 191 May Lane, Kings Heath, Birmingham B14 4AW. Currently JRT plan to offer three stages of tune for all r.h.d. Opels save the six-cylinder models, and the kits should be on the market early next year.

Because the Manta uses the same 1.9 engine as the GT, it's likely that the first proper r.h.d. British kits (which have been developed in part-association with Piper Engineering in Kent) will be for that likeable four-seater coupé. As with all Opel GTs our test car was a l.h.d. machine in standard trim, save the engine parts and 5½-inch alloy wheels, which were shod with 165-section Goodyear GP radial-ply tyres.

The stage 1 engine kit is said to boost power to 107 b.h.p. nett (compared to 90 b.h.p. in standard form), thanks to redesigned inlet and exhaust manifolding, reprofiled camshaft and single Weber 40 DCOE carburetter. The GT also had an aluminium rocker box cover, but we have excluded this non-productive item from the total cost of conversion parts at £78.50. Breaking that price down into individual items, we found that the inlet and exhaust manifolding came to £33.50 (£13.50 and £20 respectively) whilst the carburetter cost £31, assorted gaskets and linkages £2 and the deeper breathing cam-shaft £12. The higher stages of tune include a cylinder head porting modification (stage 2 at an extra charge of £39.50) rated at 123 b.h.p., and a final fling, claimed to provide 140 b.h.p., with a big valve cylinder-head and twin double-choke Weber carburetters. The "top of the line" kit costs £168.50, excluding any fitting charges.

General Motor's public relations people in London have taken a keen interest in what Rhodes has been doing (the factory approve many Steinmetz kits in Germany, as they are required to do by law) and it was through the usual chain of Opel UK command that we obtained our car. Opel dealers will further act as fitting agents for the Rhodes parts, if all goes to plan. So we left an extremely dirty Ford sitting amongst the gleaming pack of Opels at Mendells in Golders Green, and set about re-aquainting ourselves with the very comfortable and sporting cockpit that comes as part of the GT coupé.

Although our car was as shiny as the rest of the press demonstrators, we couldn't help noting that the wheel arches bore evidence of the difficulties of precisely judging width: twin racing mirrors were provided and there are rear side windows. But one still needs to be cautious about city driving, especially when swopping traffic lanes—the mirrors tend to show the accident after it's happened! Incidentally both the aeroflow grille and on/off air switch fell off on our way home, so not all German cars are faultlessly finished in comparison to British home products, some of which even cover a full test week without scattering control knobs and other minor components about the interior.

At first the automatic choke of the GT defeated our early morning starting attempts, for it's imperative that the throttle is not dabbed further after the first depression to ensure that the auto choke is engaged. The three-speed GM Strasbourg automatic gearbox tends to provide somewhat lumpy changes at first, but we overcame this slight inconvenience by using second gear hold until both transmission and engine were slightly warm. A sporting air-cleaner for the Weber carburetter betrays the conversion by hissing at low speed and gurgling heartily beyond 3,500 r.p.m. Tickover speed was a creditable 500 r.p.m. for most of the test; again it was important not to blip the throttle at all, otherwise the engine would stall upon releasing the accelerator. Pushing the efficient central auto lever into neutral increased tickover to 12-1500 r.p.m., which partially explains our traffic-ridden fuel consumption figures in the 18 m.p.g. bracket.

Somehow the demonstration bookings for this Steinmetz GT were scrambled up before we could take performance figures on a fifth wheel, but some quick runs utilising the stop-watch corrected speedometer indicated a 0-60 m.p.h. time of 10 seconds, which is not a shameful performance for a steel bodied 2-seater with automatic transmission, and is more than enough to deal with GT-badged saloon cars, unless you count the 3-litre Capri which took slightly over 10 seconds to reach 60 m.p.h. before the engine was uprated for this year's Motor Show.

What really did impress was the way in which the speedometer would swivel round to indicate 90 m.p.h. if one accelerated from 50 m.p.h. or so in top. We held the car easily at this velocity with three people crammed on board and found that the excellent body shape allows one to use half throttle at anything up to 90 m.p.h., with consequent benefit to fuel consumption figures. Flat out we managed 120 m.p.h. according to the speedometer with ridiculous ease (though under favourable circumstances) as the car just seems to cut through the air and become better mannered the faster one goes.

In town the automatic transmission was dispensed with in brisk accelerating stints, because it always engages third as one slows up for the next set of traffic lights, and the driver felt the brakes should be preserved for something better than a heated demise on the Marylebone road. The combination of good disc/drum braking, and

MOTOR SPORT, DECEMBER 1971

CONTINUED ON PAGE 79

OPEL GT

Story: L J K Setright

On the second day of practice for the Race of Champions I was proceeding in the course of my duties down the A20 when I observed two Opel GT coupés travelling in the same direction towards Brands Hatch at a fast speed. I accelerated my motorcycle and drew level with the foremost of them, when by keeping station with it I found it to be maintaining a speed of 100mph...

IT WAS FAIRLY CONclusive evidence that the Opel GT is a genuine quick car, for the two of them were close together and riding very steadily. It was also evidence that the cars are perhaps not as rare as I had thought them to be in Britain; or was it for safety's sake that they only ventured out, like policemen in tough districts, by two and two? Or perhaps it was a GM works outing? I had no time to find out, but accelerated the Bonneville away; and when I had reached the Hatch and parked the Triumph next to Denis Jenkinson's Norton I had forgotten all about GM and Opel. Recollection came a few days later; and after pursuing my enquiries with some tenacity and no little impertinence I found myself the delighted tenant of a nice Opel GT for a few days. To heighten my interest, it happened to have an automatic transmission rather than the usual dogbox, and thus gave me valuable experience with the new Strasbourg transmission that GM have designed specifically for the smaller cars such as Opel and Vauxhall build.

There are many who think that an automatic transmission is anathematic to any sports car. Sometimes I feel the same way, but I am tending increasingly to think that it has its place in any truly *modern* sports car. Until recently there has not been a conventional torque converter transmission of the size and quality necessary, all the really good ones such as Chrysler's Torqueflite being much too big. This new GM device promises to fill the noticeable gap. It is certainly small enough, and reasonably light (20lb less than the Powerglide of their Rekord and Victor cars); and because it is a divided design, with the gearbox housing separate and distinct from the converter enclosure, the basic gearbox can be combined with whatever converter seems appropriate to the duty, even for something as small as the Viva which was not well served by its previous slushpump. As for the quality of the Strasbourg transmission, I consider it up to the best of current standards. Rarely does it jerk (though its character changes perceptibly as it gets warm), and the matching of the torque converter to the Opel's engine and to the gear ratios and their control mechanism is obviously excellent: a half-throttle mid-range change may be accompanied by a variation in engine rev/min of less than 500, a sign that the necessary sums have been done very carefully. Even without looking at the tachometer, the ineffable smoothness of acceleration through the gears is proof of the care taken to ensure passage from one ratio to another at just the right time to make the transition from one tractive effort curve to the next. This shows up particularly in the full-throttle change from intermediate to top gear: it takes place surprisingly early, at 65mph—but produces a faster standing quarter-mile than when the inhibitor lever is used to hold intermediate to higher speeds such as 70 or 75.

Smooth changes such as these are vital in a sports car, where there is a strong temptation to use lots of throttle in improbable places such as the apex of a slippery corner. Violent changes in tractive effort at times such as these can have the car breaking traction and its driver's neck in rapid succession. The Opel GT has the precise and immediately responsive handling that makes very fast cornering come naturally, and any inadequacies in the automatic ▶

CONTINUED ON PAGE 78

photography: John Perkins

OPELS
WITH HAIRS ON THEIR CHESTS

AS USUAL, WE MADE AN effort during the Geneva show to drive some of the more interesting new production cars as well as to make arrangements for getting to grips with the various exotica later in the summer. First we trooped along to the luxurious l'Reserve motel on the outskirts of the city, overlooking the lake, for a morning with the Opel people, who in their usual efficient manner had assembled a whole field-full of cars for us to try. Then the following day we ventured into the western suburbs to wait upon Messrs Fiat, who had declined to provide a 130 coupé on the reasonable grounds that the only ones in existence were on show in the main hall but who had produced one of the new 128 Rally saloons for assessment.

The first of two Opels we tried was the GT/J, a simplified version of the pretty and in many ways surprisingly sophisticated GT coupé. This had caused something of a sensation at the show because of its unexpectedly low price of less than £1200 on the Swiss market, as against £1000 for the far less powerful and stylish Spitfire in which we had made the trip from London (story next time). The GT/J is undoubtedly a bit of a boy-racer machine and Opel are quite blatant about its intended appeal to people in their early 20s, but it is not all p**s and wind. The engine, for example, is the larger 1.9litre unit in exactly the same tune as for the more expensive version, with a claimed 90bhp DIN at 5100rpm and 115lb ft of torque (SAE, as we have forgotten how to translate the DIN figure from kpm) between 2800 and 3400rpm. The mechanical sophistication throughout is unaltered, but there are some spartan touches to the body as well as a very garish paint job based on some nice background colours, with matt black for what would otherwise be the brightwork and a wealth of patches and stripes.

Most important of the sacrifices are the rear quarter windows, which are now fixed, and the padding in the rear compartment, so that this is now rendered distinctly uncomfortable for any third party, however small. It is useful for luggage, though one must still shift everything out in order to get at the spare wheel in the tail as there is no outside boot lid. Most of the subsidiary instruments have gone, too, replaced by warning lights, and so have the clock, the electric washer/wiper control and most of the pseudoluxurious crap that nobody ever misses anyhow. The interior still looks and feels functional, reminding us of, for example, an Alfa GT Junior (to which the name in any case presumably alludes) in the way rubber can replace carpet for the floors without detracting too much from the general atmosphere. Obviously, what matters in these cases is not how much you leave out but what you leave in, and Opel seem to have got the blend just about right everywhere.

On the road, the new model handles just as well as a normal GT1900, which is to say very well indeed. Although relatively narrow, the chassis is cleverly conceived, with the engine set well back and all of the weight contained within the wheelbase. This gives a relatively low polar moment of inertia, so that the GT handles rather like one of the old TR sports cars—neutral to start with, giving way to definite oversteer which because of the light back end and well-located axle (coils, trailing links, Panhard rod) is easily controllable on the throttle. It is much more stable at speed than any Triumph, however, and the aerodynamics are obviously as good as they look, for on the run back into Geneva down the Lausanne motorway we got up to an indicated 120mph down a slight slope and managed to hold it.

Performance generally is brisk without being exciting, which is just as well as if there were any more one could do without the oversteer. We got speeds in the gears, indicated, of 32, 57, 87 and 120mph. The only obvious disadvantage of this version by comparison with the costlier one is that it is a lot noisier. Wind roar is still completely absent, but a good deal of the scuttle sound deadening has apparently been left out and in consequence too much engine and transmission noise finds its way to the cockpit. This is enough to make the car unacceptable to many more sophisticated users, but then think what one would do with some touch-up spray and a silent travel kit....

In fact if we were Opel Great Britain we would be thinking seriously about having this model built in right-hand-drive form, modifying it as above on arrival and offering it at something like £1400 in opposition to the outdated home-market sports cars.

Uninspired Ascona

Meanwhile, back at the motel, there was another new car to try, namely the latest and greatest in the Ascona range into which Mike Twite gave you an exciting insight last time. The 19SR de luxe differs from other Asconas not only in the size of its engine but also in the provision of two separate accessory or trim packs after the fashion of current Fords. The engine is in effect the same as in the GT/J about which we have just been talking, a high-cam five-bearing inline four of 1897cc, considerably oversquare at 93 by 69.8mm and producing an alleged 90bhp DIN, with the aid of a dual-throat downdraught Solex carburetor. Running gear is unchanged from the Ascona 16 as tested last time apart from the wheels, which are part of the GS pack, with 5.5 rims and Rostyle-type centres. The example we drove had optional, even wider tyres—185mm 70-series SR 13 Uniroyals with a widely spaced sawtooth tread pattern, the appearance of which transformed an otherwise very dull car into something with the stance almost of a BMW. Brilliant yellow paintwork heightened the impression, with a matt grille and the usual Opel stripes to go with the simulated wood dash, subsidiary instrument console, so-called sports steering wheel (not bad, but not sports) and prominent tachometer, all GS equipment.

We found the result disappointing to drive and couldn't quite work out why, for a very similarly equipped Manta which we had driven in Britain some time earlier (the Manta is identical to the Ascona apart from the outer and some inner body panels) had impressed us with the precision of its handling. This car seemed to lack any incentive to get a move on, and although it would go round corners like crazy in the dry one just didn't feel inspired to take advantage of its abilities. The engine may have had something to do with it, for it felt even less punchy than the one in the GT/J (the Ascona is only 11lb heavier at 2127lb ready to go) and throttle response was hardly instantaneous. Even the normally responsive rack and pinion steering felt dead, possibly due to the ultra-wide wheels, and the ride—much stiffened, by the feel of it, from the normal specification—was diabolical at low speeds though reasonable above about 50mph.

Opel claim 0 to 60mph in 13sec for this car, which is all right. There is no doubt that it harbours a good deal of tuning potential, and the presence of a really fierce Steinmetz-modified racing example at the show bears this out, but as supplied by the factory even with all these options (we hate to think what a UK price might be, with the much less hairy 16 standing at £1262) it is still a very ordinary little tin box.

Opel's GT/J (above) is a stripped, cheaper version of normal GT but retains all mechanical goodies. After initial-neutral steer GT/J gets twitchy in the tail (below left). Its seats are not specially comfortable (below right) compared with those of Ascona 19SR (below right), which has wide wheels (below far right) and Manta Rally engine specs.

CONTINUED FROM PAGE 74

transmission would soon be apparent. In fact the only thing to temper the driver's zeal on wet corners is a tendency for the front wheels to sidle outwards unless a little power is fed to the rear ones. Again, it is a tribute to the GT's steering that the slightest variation in tyre behaviour can be felt, though the steering is light enough for none to find it a burden.

Considering that most of the suspension and running gear, and a good portion of the chassis, come from other more humdrum cars in the Opel range, it is astonishing that the GT handles so well. To some extent the progressive springing is responsible, keeping the wheels well in contact with the road and holding the body steady. The surely futile little 1100cc version (not sold in Britain) may not be quite as good, for it does not enjoy the nitrogen-filled dampers of the 1900. Nor are its tyres as substantial as the 165HR13 Goodyear G800 radials that graced the five-inch rims of my 1900. These tyres seemed to suit it admirably, betraying it only by spinning during sprint starts on shiny wet British roads.

Only beyond 100mph does the suspension of the GT begin to lack conviction. If the road is not really smooth the ride becomes joggly, the steering begins to feel confused, and there is a general suspicion that things are getting out of shape. This is not a matter of great concern because one does not often exceed 100 in this car, even though the high 90s come up regularly. Things may be different with the dogbox replacing the automatic transmission, for there is undoubtedly more power lost through the latter, and it is only power that this car lacks. GM claim 102bhp SAE for the 1900 GT engine, but how much gets through to the road I would not like to guess. At any rate this car needed a very long run to get past 105mph, allowing for the slightly optimistic speedometer.

Certainly the body shape is in no way to blame. The frontal area is low—15.8sq ft—and so is the fuel consumption even when driven hard, as this car was for most of the 500 miles I did in it; so the drag coefficient must be as low as the look of the thing suggests. The retracted headlamps, the rounded windows set almost flush in their frames, the clean nasal entry and gentle contours must all help in penetration; and if my car had not been decked out with so many external mirrors to create parasitic drag (excusable because the GT is only made as a left-handed car, even for this right-handed country) I fancy it could have gone appreciably faster. Even so, it was good: wind noise went unnoticed, and other noises were subdued enough for the siffles to have been heard if they were sounding.

The docked tail must be effective, too, not only in reducing induced drag but also in keeping the centre of pressure well to the rear. This is essential in a streamliner, especially when its engine is set well back to assist balance and handling; otherwise crosswind instability can be very trying. In this respect the Opel is superb, and I shall not soon forget how well it proved itself. I had taken it west to look at Tintern Abbey on a day that had everything—rain, snow, hail, shine, fair winds and foul—and on the return trip was stooging peacefully along the M4 just east of the Severn when the car was suddenly smacked by a tremendous gust as it emerged from the lee of something big. I had never thought Boreas so horny-handed: it was a fearful blow, so sharp as to wrench the steering wheel out of my hands. I was going very fast at the time, and the consequences could have been nasty; but thanks be to God and Adam Opel Aktiengesellschaft, the car just jinked a little to the left and then sailed serenely on, straight down the road. My admiration for this tireless mile-gobbler soared.

In London again, I found one of its few faults. The headlamps are retracted and deployed mechanically, the driver heaving or hauling on a lever ahead of the gearstick. As he does so, the headlamp housings rotate about a longitudinal axis. The system has certain advantages: its action is neither helped nor hindered by wind pressure, the lamps are in satisfactory alignment as soon as they appear, they cannot droop through pneumatic depreciation nor stay down because of electrical slump. In fact the method is the most reliable imaginable; but it is noisy. There is an enormous clonk as the lamp housings slam into either position—and if you try to do it gently they do not locate positively enough, leaving a tell-tale light glowering at you from the facia. The noise can be embarrassing in traffic, even though no real harm is done: it sounds just like the coming together of two portions of car bodywork (which indeed it is) and everyone looks round to see who has hit what. When the fellow in the car ahead of you hops out to come and inspect the damage, only to retire baffled by the sharp little sportster two or three feet clear of his stern, you may be amused—but not if you cannot afford the delay.

The only other fault lay in the seat harnesses, which were unsuitable; but this is easily set right. Otherwise the interior was as convincing as the exterior, as good-looking and as efficient. The seats were very good even after hours of occupation, the driving position was excellent, the instrumentation and switchgear fair to good, and the quality of the trim quite impressive until one remembers that this is after all a fairly expensive car. In Britain the price starts at £2067, and you will of course insist on a laminated front glass and electrically heated back one, adding another £34. If you want the Strasbourg automatic it will cost another £137. There, though, is an end to it: Opel's list of their principal factory-fitted options runs to thirty items, but the rest may be ignored, for the 1900 GT has them all as standard or else (in a couple of cases such as rubber flooring and roof rack) is better off without them. Perhaps the only thing it wants is more power; the performance is quite brisk as it is, but the car feels as though it could use more if more were available. That is usually a good sign. ●

Despite typically American thin, hard seats and the narrow-waisted styling which gives minimal arm room, the Opel GT is surprisingly comfortable to drive (left). The 1900 engine gives this light coupé very good performance even with the automatic gearbox (right)

OPEL GT—*continued from page* 66

BRIEF SPECIFICATION OF OPEL GT—1.9-LITRE

Wheelbase : 7 ft. 11.7 in. (2,431 mm.).
Track, front : 4 ft. 1.4 in. (1,254 mm.).
Track, rear : 4 ft. 2.6 in. (1,284 mm.).
Overall length : 13 ft. 5.9 in. (4,113 mm.).
Overall width : 5 ft. 2.2 in. (1,580 mm.).
Overall height : 4 ft. 0.2 in. (1,225 mm.).
Kerb weight : 18.9 cwt. (960 kg.).
Wheels : 5J × 13.
Tyres : 165 HR 13—G800.
Steering : Rack and pinion.
Suspension : Front : transverse leaf i.f.s.; *rear :* rigid axle, coil springs.
Bore and stroke : 93 × 69.8 mm.
Capacity : 1,897 c.c.
SAE horsepower : 102 at 5,400 r.p.m.
Max. r.p.m. : 6,000.
Compression ratio : 9.5 to 1.
Main bearings : Five.
Camshaft in head : Chain driven.
Valves : Side by side operated by rockers from o.h.c.—valves at angle to centre-line of engine.
Carburetter : Single 32 DIDTA-4 Solex, double-choke.

the nose about their fore and aft axes and the headlamps appear. No worries about coming round slowly, out of phase, or on half-cock, there they are with a resounding "clang", and a pull on the lever rotates them back out of sight. When in use the very rigid indicator stalk operates the dip mechanism on the pull-back, positive-stop system, but the cut-off on dip is a bit lethal when travelling fast at night. The disc front and drum rear brakes, with dual servo, would avoid any "blackness" embarrassment and are very powerful, with a nice progressive feel, so that the harder you push the faster you stop. At maximum speed, which is in the 110-112-m.p.h. area, the car is splendidly stable and you feel you could use quite a lot more speed, and the body shape gives a very low wind-noise factor, its penetration obviously being very good.

It is described as the Opel GT two-seater sports car, and that is what it is, for the tail is filled with the 12-gallon fuel tank, the spare wheel and the jack, so that luggage space is at a minimum. In every way it gives you a confident feeling that it is the product of a serious Research and Development department and test facility, coupled with a giant production plant, in opposition to a car built by enthusiasts from proprietary parts. If the Opel GT is an attempt to instil some sporting character into the name of General Motors, and in particular that of Adam Opel, it has certainly succeeded. In Great Britain it is handled by General Motors Ltd., of 23, Buckingham Gate, London, SW1, and the basic price is £2,057 4s. 9d. with the 1.9-litre engine, and £2,227 1s. 6d. in the form tested, with automatic transmission, fatter tyres and heated rear window.—D. S. J.

CONTINUED FROM PAGE 73

reliable cornering manners via transverse leaf i.f.s. and superbly restrained live rear axle, was put to excellent use in conjunction with the conversion as soon as we slipped off along a country lane. The automatic lever instantly selects any of the three ratios, providing that the driver has assesed his speed correctly, and we found the 45 m.p.h. first hold—marked L—and S for the second slot, allowing 70 m.p.h., were very useful aids indeed. Maximum engine speed is 6,150 r.p.m. as standard, a yellow warning band commencing at 5,800 r.p.m. and we stuck to this figure in the absence of any instruction to the contrary, besides which the Opel four isn't really fond of 6,000 r.p.m.

Overall we left the car profoundly impressed by its ride and handling (again) and freshly thoughtful about it's remarkable cruising ability, for our rough check had shown only a small percentage error from the speedometer, and a genuine 110 m.p.h. can be maintained in true GT style with the Steinmetz kit, with a small power surplus in hand. It is a pity Opel do not market right-hand drive GTs.

Obviously JRT and their sub-contractors will have to apply themselves vigorously to obtaining the best traffic manners commensurate with city fuel economy. The power potential is already evident and could provide a welcome boost for Manta owners who dislike reading Ford bootlid badges. Much more exciting though are the tales from a French journalist colleague who has tried a 2800 Commodore with Steinmetz Road and Rally stage tune, yielding 180 b.h.p. DIN at 6,400 r.p.m. (compared to 145 b.h.p. in production form). This triple Solex 40-mm carburated device managed 131 m.p.h. with appropriately startling acceleration. In fact that road-going Commodore should equate approximately with the German Capri RS2600 and the Broadspeed/Super Speed/Raceproved-Weslake/Willment-tuned British 3-litre Capris. On the track Steinmetz has also tried to keep ahead of Ford Germany, designing his own crossflow cylinder head and ensuring that all the right parts were homologated for Gp2, but his hoped for 300 b.h.p. was never handled well enough, or sufficiently reliable without dry sump equipment, to catch the smooth Ford Germany Capris driven by a number of hot shoes.

Good luck Mr. Rhodes, may we also hope to see that famous tyre smoking approach demonstrated on the race tracks again next year, only this time with a bulky Commodore held at right angles to the track, instead of a Cooper S?

Comparison R&T Road Test

Comparing the Datsun 240Z, Fiat 124 Sports, Opel GT, MGB GT and Triumph GT6—a closer contest than we expected

WHAT DOES ONE get when he buys $3500 worth of Grand Touring car? Generally, a small, light closed car, adequately but not spectacularly powered, a cut above the average in handling and braking, offering a measure of comfort not found in an open sports car. This month we take a comparative look at five such cars. None of them is a new model, all having been tested before by R&T in separate road tests. There have been detail changes in all of them since we last tested them although their basic character has not changed; still there's nothing like getting a group of cars together, taking a journey in them and comparing them nose-to-nose. We're always surprised at how much we learn in a comparison test and the reader may be surprised at some of the results of this one. We were.

As the General Data table shows, the list prices of the five cars all cluster around $3500. And each car is reasonably complete at the basic price—you don't need to pay extra for adjustable seats, a cigarette lighter or a 4-speed gearbox. All are front-engine, rear-drive cars, all have inline piston engines, all have at least two disc brakes out of four, and all weigh between 2000 and 2500 lb; one is from Japan, one from Italy, one from Germany and two from England. And they are all very different from each other.

The Datsun 240Z set U.S. motoring on its ear when it appeared in early 1970. It seemed too good to be true: a really fast, good-handling and good-looking coupe with great refinement and extensive standard equipment—all at what seemed an incredibly low price. Strictly a 2-seater but a roomy one, it is the longest car of the group though not the widest, the heaviest by a small margin, and by leaps and bounds the most powerful with 150 bhp from its 2.4-liter overhead-cam 6-cyl engine. It has independent suspension all around (by struts and coils at both ends) and a combination of disc front brakes and drums rear. Its styling is professional and up-to-date if a bit "pop culture," and that the 240Z is an exciting package cannot be denied.

In fact, it is so exciting that it has generated a supply-and-demand situation virtually unprecedented in the U.S. The waiting list for delivery on one is as high as six months, even though it has been over a year since the model was put on sale. Datsun programmed a supply of 1600 per month for the U.S. and they're now getting over 2500, but they have found that the demand is for about 4000 per month. Interesting, for we asserted in the 240Z introduction story that a domestic carmaker probably could build an equivalent car for the same price ($3600) in quantities of 50,000 per year or so. Anyway, Datsun stopped production on the 1600 roadster to make way for more 240Zs but the supply is still far behind. In some areas, for instance, dealers are telling customers they can't get the cars without lots of optional equipment—wide alloy wheels, air conditioning, etc., and getting away with it. And the *Kelley Blue Book* retail value for a used 1970 240Z is over $4000!

So—though the list price of a 240Z may be only $3596—one may or may not be able to get one for that price. If you can find a dealer who will sell you one for that price you'll probably have to wait months to get it. We must give Datsun the credit for producing such a package at such a reasonable price but we have to caution the reader that Datsun's price may not be the dealer's price. The laws of supply and demand still work, list prices or no.

Standard equipment on the 240Z includes a signal-seeking AM radio (with electrically powered antenna) and a heated rear window. Our test car had no options except a set of wide wheels (14 x 5½, part no. 40300E4600, $13.50 each) that can be obtained from Datsun dealers but are not installed at the factory; for the price table we estimate a total charge of $74 for the wheels and installation.

The Fiat 124 Sport Coupe has been updated this year with a smooth new front-end look and a 1608-cc engine of longer stroke than the older 1438-cc unit. It's not any faster than before but the larger engine is stronger in the middle ranges (so that less gearshifting is required to maintain a given pace) and both smoother and quieter. The Fiat is almost as long as the Datsun and somewhat wider; it is the only car in this group to offer a real rear seat and it's so real a seat that the car can be compared to some small sedans. It has several engineering distinctions: the only twincam engine in the group (the two camshafts are driven by a single toothed belt), the only 5-speed gearbox and the only 4-wheel disc brakes. It's the only car in the group with a separate, lockable trunk where one can hide valuables; three of the others have "tailgates" and one a cockpit luggage area. And its luggage capacity is the largest in the group.

The 124 Coupe test car, working from a POE price of $3292 (the lowest in the group), had the nice Cromodora alloy wheels, which cost only $135 for the set, a rather poor AM/FM radio (one shouldn't expect much at $85) and add-on chrome side strips and luggage rack; the last two items aren't included in our price tabulation.

British Leyland's MGB GT has been around since 1966

THE $3500 GT

and is based on the MGB roadster introduced in 1962. It's a classic British sports car but with a nicely designed fixed roof. The only one of the group attempting to be a 2+2, it has a very small bench seat behind the two main ones; this is big enough for small children only, and it can be folded to extend the luggage area. Surprisingly, the B is nearly as heavy as the Datsun, and with only 92 bhp from its pushrod four it is the slowest of the lot. Its front suspension is conventional—unequal-length A-arms and coil springs—and rear suspension is by the simplest means possible, a live axle on leaf springs.

The MGB GT comes with radial tires as standard equipment; styled steel wheels are standard, wire wheels (either painted or chrome) optional. Our car had the optional overdrive ($165), the familiar Laycock-de Normanville hydraulic unit that shifts in or out at a flick of a stalk on the steering column to give a 0.802:1 reduction. Thus 4th gear becomes 3.14:1 overall instead of 3.91:1 and the MG becomes the longest-legged of the lot. The overdrive works on 3rd gear also.

Opel's GT is based on the Kadett chassis, so it actually isn't as up-to-date in that department as its bulkier cousin the 1900 Rallye. Its front suspension is odd, based on a transverse leaf spring and one set of lateral control arms, but at the rear everything is shipshape with a live axle located by trailing arms and Panhard rod and sprung by coils. Its all-steel unit body (many people assume it's plastic because it looks so much like a Corvette) is made in France to a very high standard of finish and for 1971 it is available only with the 1.9-liter engine rather than both the 1.1 and 1.9. The engine has taken a power cut, though, to satisfy the politicians; bringing the compression ratio down from 9.0 to 7.6 so it could run on 91-octane fuel has reduced power from 102 to 90 bhp. Performance has suffered (0-60 time is up from 10.8 to 11.9 sec) but the GT still holds its own in the class. Amazingly, the GT has had a price cut too—it's now nearly $200 cheaper than it was in 1969 with the 1.9 engine.

The Opel GT is strictly a 2-seater: luggage is carried on a flat floor behind the passengers and loaded in through the car doors. The spare tire lives behind a vinyl partition just aft of the luggage area. Thus it has the least convenient luggage accommodation, though not the least capacious. Other distinctions for the Opel: it's the lightest and most economical of fuel in the group and its brakes can stop it in the shortest distance. Our test car had but one extra, a good AM

PHOTOS BY GORDON CHITTENDEN

$3500 GT

Datsun's engine is the largest and its instruments and controls the best.

Fiat has the only dohc engine and excellent instrumentation.

Opel's engine has generous displacement and unusual high-cam design.

MG has time-proven pushrod engine, safety-modified instrument panel.

GT6's smooth engine is very accessible, its wood-panelled dash luxurious.

radio at approximately $75 (dealer installed); like the Datsun, it can be ordered with a 3-speed automatic transmission.

Before the 240Z appeared the Triumph GT6 was the only car in its class with a 6-cyl engine, but now its 95-bhp, 2-liter six is no longer a big attraction. It's the smallest car of the group and only insignificantly heavier than the Opel; these points relate to the fact that it's derived from Triumph's smallest, lightest sports car, the Spitfire. It shares basic chassis structure (a backbone frame with separate body), front suspension and body structure from the beltline down with the open-bodied, 4-cyl Spitfire though the Spitfire's swing-axle rear suspension is replaced in the GT6 by a more satisfactory unequal-arm linkage for the rear wheels. The GT6 has the tightest interior dimensions of the group but not the smallest luggage capacity, and its luggage area is loaded easily through a tailgate as in the Datsun and MG.

This year the GT6 has been freshly restyled on the outside, and if some staff members noted that the rear end is reminiscent of the old Sunbeam Harrington coupe, all agreed that the car is better looking than before. It has one feature that makes it uniquely maneuverable in a crowded city: a tiny turning circle made possible by front wheels that can be steered well beyond the limits of proper geometry. One can turn around in just 25 ft, 6.5 ft tighter than the next twistiest car in the group, the Datsun.

Standard equipment at the GT6's basic price of $3424 includes whitestripe radial tires and a heated, tinted rear window; it has a 4-speed gearbox and an overdrive like the MG's can be ordered as an option; brakes are a disc/drum combination. Our test car had only an AM/FM radio—again a rather poor and low-priced ($100) one. A final distinctive feature of the GT6: A rich wood-panel dashboard in the traditional British manner.

As in past comparison tests, we chose a test route appropriate to the character of the cars. In this case it was a route we'd used two years ago for a group of more expensive GT cars. We left our office in Newport Beach, topped up all the cars at a nearby filling station, and drove south

GENERAL DATA: 5 MEDIUM GTS

	Datsun 240Z	Fiat 124 Sport Coupe	MG B GT	Opel GT	Triumph GT6 Mk3
Basic POE price*	$3596	$3292	$3620	$3306	$3424
Price as tested	$3745	$3562	$3823	$3409	$3674
Engine position/driven wheels	f/r	f/r	f/r	f/r	f/r
Chassis type	unit	unit	unit	unit	separate
Brake type	disc/drum	disc	disc/drum	disc/drum	disc/drum
Swept area, sq in./ton	233	227	227	222	209
Suspension, front	ind. coil	ind. coil	ind. coil	ind. leaf	ind. coil
rear	ind. coil	live coil	live leaf	live coil	ind. leaf
Standard tires	175-14 rad	165-13 rad	165-14 rad	165-13 bias	155-13 rad
Steering turns, lock-to-lock	3.5	2.7	2.9	3.0	4.5
Steering index	1.10	0.99	0.93	0.99	1.13
Fuel tank capacity, gal	15.9	11.8	12.0	13.2	11.7

*POE prices vary slightly for east, west and Gulf ports
As-tested prices include: for Datsun, 14 x 5½-in. wheels and installation; for Fiat, alloy wheels, AM/FM radio; for MG, overdrive; for Opel, AM radio; for Triumph, AM/FM radio. All as-tested prices include charge for preparation at dealer.

Opel's separate lap-shoulder belts work but are a mess.

on the Coast Highway through Corona del Mar and Laguna Beach (stoplights galore) to Dana Point, where we turned inland to connect with Ortega Highway, California 74. This highway, a nicely surfaced, 2-lane route with a delightful variety of turns and hills as it winds through some of Southern California's finest country, was lightly traveled and the weather was beautiful; it was easily the highlight of our trip. At Lake Elsinore we connected with route 71 toward Temecula; after Temecula, route S16 to Pala, 76 to Santa Ysabel and 78 to the little town of Julian high in the Laguna Mountains, where we stopped for lunch. Then on down 78, another wonderfully twisty road, into the Anza-Borrego desert, out across the desert at speeds dictated by road conditions and car capability rather than artificial limits, up through the Joshua Tree National Monument, and over clear, generally straight back roads to our overnight stop at Victorville, from where we freewayed it back to Newport Beach the next morning. In all, a rich and varied 500 miles of motoring in which we found out all about the five GTs.

Back at the office we set about scoring the cars. Each driver was given a score sheet on which he could rate each car on 15 different aspects of behavior, such as handling, ride, quietness, braking, steering, gearbox, engine, controls, seating, ventilation and heating, vision, finish, luggage accommodation and so forth. All these categories could be scored on a 1 to 10 basis, 10 being the score for a topnotch performance and 1 being the lowest score possible. These scorings were then totaled for each driver and for the entire group to get an overall rating score for each car.

In addition, each driver was asked to rank the cars in the order of his personal preference—disregarding, if need be, his separate ratings of the car's various aspects. Here's how the ratings turned out:

The Datsun scored the highest point total. In individual driver scoring, it garnered the highest number of points from four of the five drivers, and three of the five drivers rated it their personal favorite.

Next came the Fiat, and here we emerged somewhat surprised. It had been generally anticipated that the Datsun would win by a large margin, but not so. The Fiat tallied an impressive score, little less than the Z; one driver gave it more points than he gave the Datsun, and the same driver gave it his personal nod.

Then the Opel. It was a clear step below the Datsun and Fiat but clearly not in the doldrums. One driver rated it his personal favorite, though in scoring he had given the Datsun more points.

In total points the MG was not as far below the Opel as the Opel was below the leaders, and there was no unanimity in the personal ratings of the MG by the various drivers: two rated it third, two fourth and one last. But these ratings averaged a 4th-place finish just as clearly as the points score indicated; in fact, averaging the "position" of each car over the five drivers' listed orders of preference, the cars stacked up the same way: Datsun, Fiat, Opel, MG, Triumph. Which brings us to the Triumph: It came in last, but not far behind the MG and was ranked last by four of the drivers on their personal ratings. The one driver that ranked it 4th instead of 5th also gave it more performance points, so it had a clear attraction for him. Now let's look

GENERAL SPECIFICATIONS: 5 MEDIUM GTS

	Datsun 240Z	Fiat 124 Sport Coupe	MG B GT	Opel GT	Triumph GT6 Mk3
Curb weight, lb	2355	2220	2345	2110	2115
Test weight, lb	2770	2620	2725	2500	2490
Distribution, f/r, %	51/49	55/45	49/51	54/46	54/46
Wheelbase, in	90.7	95.3	91.0	95.7	83.0
Track, f/r	53.3/53.0	53.0/51.8	49.3/49.3	49.4/50.6	49.0/49.0
Length	162.8	162.3	152.7	161.9	149.0
Width	64.1	65.8	59.9	62.2	58.5
Height	50.6	52.8	49.4	48.2	47.0
Luggage capacity, cu ft	8.5	9.6	6.3	6.6	6.6

MG's tailgate-loading luggage area is handy.

JULY 1971

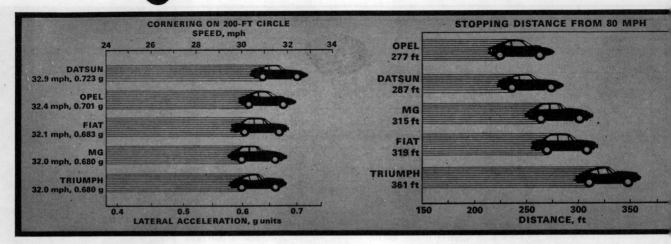

THE $3500 GT

at the cars in detail and in the order of their ranking.

Datsun 240Z

IN GENERAL the 240Z lives up to its promising specification. Its generous-size engine delivers the smoothest, quietest and dramatically the most powerful performance in the group; it is an engine that pulls strongly from low speeds, runs silently at high cruising speeds and continues to be impressive right up to its 6500-rpm yellow line on the tachometer, thanks to improvements in the crankshaft since our earlier road test. All this performance isn't without cost; the "Z" is also the thirstiest in the group, but 21 mpg is nothing to complain about and the fuel tank is large enough to give it a cruising range to match its cruising ability.

But the Z-car has a serious problem associated with high-speed motoring. It's very sensitive to sidewinds at speed, and when traversing road undulations at high speed (as we did repeatedly on the desert highways) it requires a lot of motion at the steering wheel to keep it on course. Datsun has some chassis tuning work to do here, and in the meantime an owner can fit an undernose spoiler, available from BRE (137 Oregon St., El Segundo, Calif. 90245) for $32.

In low- to medium-speed cornering and handling, however, the 240Z shines. Surprisingly, the 5½-in. rims didn't make a difference in absolute cornering power over the standard 4½-in. ones, though they added crispness to the car's response; but it still led the group with a 0.723g cornering capability. And though its steering is not the most pleasant or accurate in the group, it is quite acceptable, never unduly heavy and certainly quick enough.

The Z's combination of front discs and rear drums, vacuum assisted, tie with the Opel's brakes for best-in-group; though the "panic" stopping distance from 80 mph takes 10 ft more than the Opel, the car stays under control a little better under these conditions and fade under hard repeated use is negligible.

Comfort and accommodation also rank high in the 240Z. It accommodates only two people, but what space for those two! The largest driver in our test crew, who is 6 ft 2 in. tall and weighs 200 lb, gave it the full 10 points on every aspect of its interior but finish—undeniably some of the materials used, notably the quilted-pattern vinyl, are less than pleasing. Controls are notably good, everything being within good reach for a typical male driver with his 3-point belt fastened (the best belt in the group, by the way), and the steering column-mounted lighting control rates special mention. On the other hand, it was the only car in the group without some provision for easy daylight headlight flashing, and the ventilation system, though it provides a good flow of outside air that can be boosted by the blower, unfortunately aims most of it right between the driver and passenger.

In sum, the Datsun 240Z's plusses are its striking looks, its effortless, strong performance, its good brakes and low-speed handling, and its comfort and equipment. On the negative side, the only serious criticism is about the high-speed stability. If you can get one for list price, or even get one with the extras *you* want, it is not only the best car in the group but the best buy.

Fiat 124 Sports Coupe

THE FIAT deserves more popularity. At nearly $200 less than the Datsun with comparable equipment, it did so well in our comparison test that it scored nearly as many points. Of course it doesn't offer the zoomy styling of the Datsun (the boxy shape that turns it into a true 4-seater doesn't allow that) nor the brilliant performance. Its 4-cyl engine, the smallest of the group, is nevertheless a most satisfying bit of machinery: quiet, very smooth for a 4-cyl (easily the best four in the group), and willing to rev happily to its 6500-rpm redline. And the 5-speed gearbox is the best gearbox in the group.

In road behavior the Fiat scores at the head of the group. Its steering is the most precise, its handling the best; it really shines at high speed in contrast to the Datsun, for it isn't blown about by sidewinds and can negotiate high-speed dips and humps without a hint of losing its composure. Only an over-eager vacuum brake booster detracts from its overall roadworthiness; our drivers always found themselves overdoing it with the brakes when first getting into the car. And though it's the only car in the group with disc brakes all around, it doesn't do anything impressive in a panic stop and the brakes squeal often when not in use. Its brakes do have the best fade resistance in the group, though.

The Fiat's driving position is, in a word, odd—and perhaps something we, as Americans, will never understand. The steering wheel is buslike in that it is less vertical than usual, and it is far away from the driver while the foot pedals are close. But the seats are good and so are the controls, which make maximum use of modern steering-column stalks to do various things. The seatbelts were installed incorrectly on the test car so that either lap or shoulder section was twisted. Ventilation is not particularly good, but ventwings in the doors make it possible to drive at moderate speeds with the door windows open and no drafts. The Fiat has the best vision outward of any car in the group, so it's a good car for city traffic.

With its good rear seat and capacious, separate trunk the 124 Coupe is far and away the most practical car of the

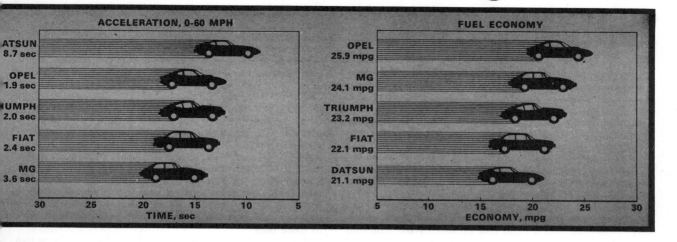

group if one needs more than 2-passenger accommodation. And surprisingly it doesn't give up a thing in sportiness for this extra measure of utility; in fact, it is *the* driver's car of the group. It is also well finished and equipped. On the minus side are its modest performance, grabby brakes and fuel economy that isn't impressive. We placed the Fiat 124 Sports Coupe as a close second in the group.

Opel GT

THIS ONE, coming in third in the group, is a crisp little package but not an impressive value for the money. Regardless of what one may say about its pseudo-Corvette styling, it is extremely well finished and put together and ranks fairly high in its driving position and control layout. It has the 2nd-best gearbox, light if not dead-accurate steering and the highest performance level of "the others" (i.e., other than the Datsun) despite its excellent fuel economy. It's the only one of the group that uses regular fuel.

The body is structurally solid and rattlefree. Vents at the dashboard ends provide a good flow of ventilating air, and longlegged gearing makes the engine (which is rather hashy sounding up through the gears) fairly quiet at speed in 4th; thus one can cruise with the windows up even in warm weather at an untiring noise level. On curvy roads the GT is amazingly well-planted and stable considering its humble origin, but in hard low-speed turns the rear axle chatters and its general handling characteristics are too much on the understeer side for maximum entertainment value. As we said in our road test of it, it may not be a really good-handling car but it is safe and predictable.

Fiat offers the only 5-speed gearbox in the group.

ENGINE & DRIVE TRAIN: 5 MEDIUM GTS

	Datsun 240Z	Fiat 124 Sport Coupe	MG B GT	Opel GT	Triumph GT6 Mk3
Engine type	L6 sohc	L4 dohc	L4 ohv	L4 sohc	L6 ohv
Bore x stroke, mm	83.0 x 73.3	80.0 x 80.0	80.3 x 89.0	93.0 x 69.8	74.7 x 76.0
Displacement, cc	2393	1608	1798	1897	1998
Bhp @ rpm	150 @ 6000	104 @ 6000	92 @ 5400	90 @ 5200	95 @ 4700
Torque @ rpm, lb-ft	148 @ 4400	94 @ 4200	110 @ 3000	111 @ 3400	117 @ 3400
Transmission	4-sp man[1]	5-sp man	4-sp man[2]	4-sp man[1]	4-sp man[2]
Standard final drive ratio	3.36:1	4.10	3.91	3.44	3.27
Engine speed @ 70 mph, rpm	3350	4000	3980	3560	3480

[1] 3-sp automatic optional   [2] overdrive optional

Datsun's performance is in a class by itself.

PERFORMANCE: 5 MEDIUM GTS

	Datsun 240Z	Fiat 124 Sport Coupe	MG B GT	Opel GT	Triumph GT6 Mk3
Lb/hp (test weight)	18.5	25.2	29.6	27.8	26.2
Top speed, mph	122	112	105	110	107
Standing ¼ mi, sec	17.1	18.6	19.6	18.4	18.6
Speed at end of ¼ mi, mph	84.5	72.5	72.0	74.0	74.5
0–60 mph, sec	8.7	12.4	13.6	11.9	12.0
Brake fade, % increase in pedal effort in six ½-g stops	10	nil	17	nil	20
Stopping distance from 80 mph, ft	287	319	315	277	361
Control in 80-mph panic stop	excellent	very good	fair	good	good
Overall brake rating	very good	good	good	very good	fair
Actual speed at in. 60 mph	61.5	56.0	58.0	56.0	58.0
Fuel economy, mpg (trip)	21.1	22.1	24.1	25.9	23.2
Cruising range on full tank, mi	335	260	290	340	270

JULY 1971

THE $3500 GT

In recognition of the value of seatbelts—and in the hope that we can influence more people to use them—we're rating the belts in each car. Opel has followed U.S. GM practice by simply fitting separate lap and shoulder belts, each with its own pushbutton buckle. These fit well once in place and the roof anchorage for the shoulder belt is far superior, for instance, to the Fiat's which is on the body side behind and slightly below (that is bad—collarbones can be injured) the shoulder. But the separate belt arrangement makes it extremely inconvenient to strap oneself in, thus making it less likely a driver or passenger is going to do it.

Another safety-related item: throne-type seats used to meet the federal government's head-restraint rule impair vision to the rear, and a blind spot created by the rear roof is a further vision problem. The GT needs more outward vision for traffic maneuvering.

But all in all, it's a pleasant if not exciting little coupe. Not a bad car at all—it's just that the Datsun and Fiat are so good.

MGB GT

WE'VE HEARD that British Leyland is simply letting the MGB run its historic course; when it can't be sold anymore they'll drop it and that's that. The car seems to bear it out. Meeting the U.S. crash-safety regulation was done by laying an ugly, add-on instrument panel over the existing one and the little bit of styling facelift has been done in a haphazard way.

It's truly a car of the past. Everywhere there's evidence of a sports car designed and built in the traditional manner—in a rather homemade way, to be blunt, in great contrast to the professional design and execution of the Datsun, Fiat and Opel.

That impression carries through on the road. The GT is heavy (nearly as heavy as the Datsun) but gets only 92 bhp from its noisy pushrod engine; so it's the slowest of the group. And a rather balky shift linkage doesn't contribute to driving fun—a surprise, because this was one area in which MGs always excelled in the past.

The optional overdrive does make the B GT a capable long-distance tourer; in OD at 70 mph it's turning 3190 rpm, vs the 3980 given in the Engine & Drive Train table for normal 4th gear. And the overdrive gives it the second-best fuel economy figure. But don't expect the MG to be quiet at speed even with overdrive; there's so much wind noise you'd think it was a roadster, not a coupe.

The B handles well enough but rides very stiffly. At the limit there is a bit of oversteer that makes it fun to toss the car around, especially on low-speed curves. The steering is heavy, but the MG has the quickest steering in the group. The brakes are about average.

Ventilation, provided by a simple flap under the dash, is ineffective compared with the best of the group, but one can at least maximize it by opening the door ventwings and/or the swing-out quarter windows.

Vision outward is quite good, and MG augments it with a curious righthand fender mirror, stuck out there all by itself. The seatbelts are of the simple Kangol variety, same as on the Fiat, and someone at the factory or distributor had also installed them wrong so that one section of the belt had to be twisted.

There's little to redeem the MGB GT, not even a low price, and we can only call it a holdover from another era.

Triumph GT6 Mk 3

THE GT6 IS almost as close to the MG in our score-giving as the Fiat to the Datsun. It rates close to the Opel in performance, and with six cylinders its engine is smoother and quieter than all but the Datsun and Fiat. If overdrive is ordered, Triumph installs a 3.89:1 final drive rather than the 3.27:1 of the test car and in this form it will be a bit quicker through the gears. But what promises to be a good open-road car (if the smooth, adequately powerful engine is any indication) turns out not to be because a drumming driveline vibration sets in at about 65 mph and stays there as speed rises.

The gearbox is stiff-shifting and the shifter's H-pattern is oddly skewed; this all takes some getting used to and perhaps the owner could adapt. In any case, the GT6 is a decently enjoyable car over a curvy road at moderate speeds, with light if not particularly quick steering and good handling response. But in ultimate cornering power it is at the bottom of the group. Don't expect the brakes to accomplish much—though their fade resistance is adequate they take a very long distance to stop the car from 80 mph. On bumpy or irregular road surfaces the GT6's backbone-plus-body structure is the least staunch in the group, creaking and rattling when the going gets rough.

There are charms to the GT6. Its interior materials look the richest of any in the group and the instrument layout is particularly handsome. The seats have been upgraded in recent years (as have the MGB's), but in the GT6 they have "throne" backs like the Opel's which are a bit restrictive for rear vision. Also like the Opel, the GT6 has a blind quarter that hampers one's ability to maneuver freely in traffic.

Though the GT6 scores esthetically and in performance over the MGB, it loses it all on comfort. Its seating is the most cramped in the group, the steering wheel is very high, and its seatbelts were next to impossible to adjust to fit anybody. Triumph has gone to greater lengths to update the GT6 than MG has the B, but it still failed to make much of an impression on R&T's five testers and had to be rated last-in-group.

ONE OF OUR five drivers commented after the trip that the Datsun should be rated separately, as it is simply a class above the rest. But when all was said and tallied, the Fiat came surprisingly close and the Opel was far from unpleasant. As for the two Britishers, we do not wish to kick dead horses and sincerely hope that England will be able to get off her duff, produce some competitive cars again and challenge the other countries. We have reason to believe that British Leyland does intend to keep building sports cars and to come up not only with new designs but to realign the product "mix" of MG, Triumph and Jaguar. One of these new products, we would predict, will be a medium-price GT replacing both the B and the GT6—one that we hope will render the choice of a good $3500 GT a bit more difficult to make.

OPEL

German-built sportster for the American market.

Built in the General Motors plant in Germany, the Opel GT and other models bearing this name are sold in the U.S. through Buick dealers. Designed as a strictly 2-seater, the sleek little 95.7-inch wheelbase sportster shares this same dimension with the Opel 1900 and Manta sedans. Styling for the GT has changed little since its introduction several years ago, and it still bears a strong resemblance from some viewing angles to the Chevy Corvette.

The body and frame, all of steel, are integral, providing strength with light weight, a good way to achieve performance with relatively small displacement engines. Exterior changes are slight, mainly in the bumpers to provide better low speed contact protection. New sport type wheels that combine bright highlighted 4-spoke pattern with dull black and the chrome of the bumpers are the only large pieces of shiny trim on this very smooth styling concept. The retractable headlights, when folded, give the relatively long hood line an even longer look, while the short and rather blunt rear end treatment gives the GT a rather bullet-like silhouette.

Powered by an all cast iron 4-cylinder engine that features a single in-head camshaft, the engine is rated at 75 SAE net horsepower at 4800 rpm. This engine is used throughout the Opel line and can be coupled to the standard 4-speed all-synchro manual transmission or the optional 3-speed automatic. The rear axle ratio of 3.44:1 is the same for either transmission.

Springing for the rather conventional suspension system is rather unusual, with a single 3-leaf transverse spring used on the front unequal length arm front suspension and coils on the solid rear axle. Shocks on the front are Opel, with gas-filled Bilsteins at the rear. Steering is by rack and pinion. Being a front engined vehicle, weight distribution without passengers is listed as 1135 pounds front, 985 pounds rear. However, with the seating well back of the center, passenger load distribution is designed to put 37.6% in front, 62.4% rear, which kind of evens out weight distribution, depending of course on the weight of the passengers.

The GT interior is strictly Gran Tourisimo, with bucket seats, center console with short stroke shift lever, and full instrumentation. The seat backs recline and retractable seat belts and shoulder strap provide the proper safety restraints. Makers of small sports type vehicles have for the most part, easily met existing safety standards, and are building and tuning the engines to conform to emission control regulations. The GT engine, with a Solex down draft carburetor, uses regular grades of fuel and has a vapor control system and an exhaust gas recirculating system, both designed to reduce exhaust emissions. Hard starting, especially with hot engines, has been somewhat of a problem with the special tuning required for proper exhaust controls, and the new GT engine has a carburetor venting system to eliminate this problem. An electric automatic choke eases cold weather starting.

The Opel GT is a comfortable and nice handling machine, designed for the non-racing type of fun driving and rallying. The Opel 1900 sedan with the same engine has made quite a mark for itself in amateur sports car racing, however, the GT owner, by and large, is very happy to do his fun driving as a tourist, not a racer. Handling is quite good, yet the ride is smooth, not choppy, and power assisted front disc brakes can provide a lot of fade free stopping under all weather conditions.

In addition to the optional automatic transmission, owners who demand the ultimate in all-weather passenger comfort can order a dealer installed air conditioner. With a full coupe top and roll up windows, plus excellent interior insulation against both sound and temperatures, only a low capacity air conditioner is required and there is not a great drain on engine power to operate it.

OPEL GT
Data in Brief

DIMENSIONS
Wheelbase	95.7 in.
Overall length	161.9 in.
Height	47.4 in.
Width	62.2 in.
Tread — front/rear	49.4/50.6
Steering type & ratio	rack & pinion
Fuel capacity	11.9 gal
Luggage capacity	5.4 cu ft
Design passenger load	2 passenger
Turning diameter	33 ft
Curb weight	2120 lbs

ENGINE — Standard
Type	Cam-in-head 4-cylinder
Horsepower	75 at 4800 rpm
Displacement	115.8 cu in.
Torque	92 lbs/ft at 2800 rpm

DRIVELINE
Transmission	4-speed all synchro manual, automatic opt.
Drive axle ratio	3.44:1

BRAKES
Front	Caliper disc
Rear	drums

SUSPENSION
Front	independent, unequal control arms, leaf spring, tubular shocks
Rear	solid axle, coil springs, control arms, tubular shocks

WHEELS & TIRES
Wheels, type & size	Steel, 5J × 13
Tires, type & size	165 × 13 Bias ply

NA — Data not available
DNA — Data not applicable

ROAD TEST/FEBRUARY 1973

ENGINE & DRIVE -- WHERE SHOULD THEY GO?

In handling and traction tests of conventional, front-drive, mid-engine and rear-engine layouts in four similar cars, we found each to have its own advantages and disadvantages.

PHOTOS JOE RUSZ

COMPARISON R&T ROAD TEST

IN THE BEGINNING it didn't much matter where the engine was, so long as it was near the driving wheels. The important thing was whether or not a car actually moved under its own power, not how fast it traveled or how well it went around corners. But as cars became more sophisticated and people became more sophisticated about cars, controversy arose concerning the inherent superiority of one design over the others. One famous manufacturer of front-wheel-drive cars argued that since no one in his right mind would put the cart before the horse, an engine should be used to pull the car instead of pushing it. These and similar arguments continue, and though automotive design practice has evolved into four basic categories—front engine/rear drive, rear engine/rear drive, front engine/front drive, and mid-engine/rear drive—there is still a general lack of agreement as to which is best.

One area in particular—vehicle stability and handling—has been the subject of some of the most heated debates. Each manufacturer can list any number of reasons why its particular design is superior. That some carmakers have engines and drive trains in more than one location in similar cars and can present equally convincing arguments for each with a straight face leads to further confusion. Better to say that each has its own unique characteristics, advantages and disadvantages.

With that general viewpoint, we set out on this test to find out just what those characteristics, advantages and disadvantages are. Our year-in, year-out experience with cars of the various mechanical arrangements led us naturally to have some preconceptions. Some of these were confirmed by our tests; others were contradicted. Before we go to our tests and their results, let's first discuss the theory and practice of the four layouts.

Front Engine/Rear Drive

DESCRIPTIONS OF the first automobiles as horseless carriages are really quite appropriate. They were literally carriages—short wheelbase, enormous ground clearance, large-diameter wheels and all—with a small engine fitted under the seat. At that time this was adequate for low-speed motoring with two or three passengers. But cars began to grow larger, especially in America, and space for six or seven passengers was demanded. Engines had to be increased in size to pull (or push) these loads and it soon became apparent that locating

Fig. 1	**TEST CAR SPECIFICATIONS**			
	Opel GT front engine/ rear drive	Saab Sonett front engine/ front drive	Porsche 914 mid-engine/ rear drive	VW Karmann Ghia rear engine/ rear drive
General:				
Curb Weight, lb	2035	1830	2145	1960
Weight Distribution (with driver), front/rear, %	55/45	60/40	47/53	42/58
Wheelbase, in.	95.7	84.6	96.5	94.5
Track, front/rear	49.4/50.6	48.5/48.5	52.4/54.0*	51.3/52.7
Length	161.9	160.0	159.4	165.0
Width	62.2	59.1	65.0	64.3
Height	47.4	47.0	48.4	52.0
Engine:				
Displacement, cc	1897	1698	1971	1584
Bhp @ rpm, net	75 @ 4800	65 @ 4700	91 @ 4900	46 @ 4000
Torque @ rpm, lb-ft	92 @ 2800	85 @ 2500	109 @ 3000	72 @ 2800
Chassis:				
Wheels	13 x 5	15 x 4½	15 x 4½*	15 x 4½
Tires (original equipment)	165-13 bias	155-15 radial	165-15 radial	6.00-15 bias
Manufacturer's recommended tire pressures, front/rear, psi	19/23	24/22	26/29	18/27
Front Suspension	Independent, Trans Leaf Spring	Independent, Coil Springs	Independent, Torsion Bars	Independent, Torsion Bars
Rear Suspension	Live Axle, Coil Springs	Beam Axle, Coil Springs	Independent, Torsion Bars	Independent, Torsion Bars

* 15 x 5½ wheels are original equipment on 914 2-liter; 15 x 4½ wheels were used for this test and gave test car these track dimensions.

Balanced weight distribution of a mid-engine layout is a disadvantage in the snow, but "dead" weight on the driven wheels can be used to get through low-traction situations.

the engine under the seat was no longer practical.

The geography of the U.S. had an influence on engine placement as well. Long, flat stretches in the midwest and the high mountains of the west required cars with greater horsepower and as metallurgy was in its infancy, increasing power meant larger and larger engines. Displacements of 500 and 600 cu in. weren't uncommon—try fitting one of those monsters under the rear seat!

The effect of racing on automotive design was also strong. Cars that were successful in racing sold well to the public and those early American racers were big, brutish affairs with enormous engines located up front. But even then designers knew something about wheel loadings and the effect of weight distribution on handling, so many of these cars were actually forward mid-engine designs with the engine several inches behind the front axle for better weight distribution.

Over the years a great wealth of know-how about designing cars with front engines and rear drive was built up, and it was cheap and convenient to incorporate a high proportion of well-tried components in each succeeding new model. So the conservative auto industry, especially in the U.S. and Britain, was reluctant to change from the established front engine and rear drive.

There are advantages other than cost and design experience to a front engine/rear drive layout. Size is a major factor. Packaging five or six passengers plus luggage is a more difficult engineering task with a rear engine design and to date it hasn't been done with a midship engine in the current sense. A front engine/rear drive car lends itself to an inexpensive independent front suspension for improvements in ride and steering. A solid axle at the rear cannot offer the first-class ride of independent designs but does have its advantages. One is low cost; the other is the elimination of camber change with its possible extra tire wear and abrupt reversals in handling characteristics.

The effect of ram air through the front grille and forward-mounted radiator makes a front engine relatively easy to cool and accessibility of commonly serviced components is generally better than with mid- or rear-engine designs. Today, safety is an obvious consideration and most designers see in the front-engine layout the only solution to providing adequate crush space (for protecting passengers in a front-end crash) at reasonable cost.

Then there's handling. A noseheavy front-engine car understeers under most conditions and engineers view this as an asset for the average driver. A driver is less likely to get into serious trouble upon entering a corner too fast if the car's front end runs wide than if its tail has a tendency to come around.

There are minuses to a front engine/rear drive layout as well. The driveshaft is a necessary evil of any front engine/rear drive car; it encroaches on passenger space and makes for uncomfortable seating for the middle passengers. An aerodynamic body shape is more difficult to achieve when a designer has to work around a bulky engine up front since it won't let the nose taper as sharply as it might. For racing applications the front-engine car is also at a disadvantage as far as driver comfort is concerned: cockpit heat is a serious concern in most big front-engine race cars. Weight transfer toward the rear when accelerating is an asset in a rear-drive car, but wheelspin can still pose a problem if the proportion of weight on the rear wheels is low. Conversely, weight transfer toward the front during braking unloads the rear wheels too much in a very nose-heavy car to fully use the four tires' friction capabilities.

Front-Wheel Drive

FRONT-WHEEL drive for road vehicles goes back long before the invention of the internal-combustion engine, the most famous ancestor being Cugnot's 3-wheel steamer of 1770. The first car to actually use fwd was patented in 1904 by an American, Walter Christie. He mounted the 4-cyl engine transversely across the frame, a disposition that has gained widespread acceptance, particularly in Europe, over the past decade.

After Christie's racing car, there was a long time before any company used the principle in a series production car. In Germany during the Thirties the low-priced Adler proved popular as did the fwd DKW, and Europeans developed an interest in fwd. Citroen has built nothing but fwd cars since 1936 and in England fwd was given a boost by various Austin and Morris models starting with the Mini. Today companies such as Fiat, Saab, Renault and Audi build fwd cars with technical and commercial success.

In the U.S. fwd is new only to the youthful and forgetful; at least five such cars have been produced in some volume. The last one before WWII was the legendary but relatively short-lived coffin-nose Cord, discontinued in 1937. The first modern American fwd design, the Oldsmobile Toronado, appeared in 1966 and was followed one year later by the related Cadillac Eldorado. Though technically interesting, these cars are successful more for their luxury appeal than the esoteric aspects of fwd.

In Europe and Japan the current trend to smaller urban-type cars has resulted in several fwd designs: Peugeot 104, Renault 5, Honda Civic and Subaru models to name just a few.

Why such interest in fwd? A fwd layout gives maximum interior passenger and luggage space with minimum outside dimensions, particularly when the engine is transverse in the chassis. It's easy to see why. Combining the engine, transmission and final drive into one unit makes for a compact power package. The driveshaft is eliminated, making a flat floor possible, and even a simple beam axle at the rear reduces trunk intrusion to a minimum. There is a drawback to this design sophistication: added cost. Contributing factors include the more costly gearing on a fwd car and the rather tricky front axle system. More money must also be spent to keep noise and vibration from the power unit out of the passenger compartment. But there are offsetting cost savings as well. A beam axle is simpler to suspend than the solid live axle of a rear-drive car (although some fwd cars have independent rear suspension). The driveshaft is also eliminated—another saving.

Low-speed winter capability of the various drivetrain configurations was tes at the Goldmine ski area, Big Bear, Calif. Each car was accelerated from rest flat section up a slope with an incline of 7-10%; the fwd Sonett performed

Front engine/front drive cars are very noseheavy, 60% or more of the total weight on the front end being common, so excessive front tire wear can be a problem. On large fwd cars a further disadvantage is incurred from the extreme forward weight bias; power steering becomes a necessity rather than a nicety.

Weight bias and the forward driven wheels combine to explain the driving characteristics that are unique to a fwd design. On a steep gradient, for example, weight transfer from the front to the rear wheels helps traction with rear-wheel drive and reduces it with fwd. But the disproportionate successes of fwd cars in rallies and ice racing provide the real answer to suggestions that they lack traction in difficult conditions. With rear-wheel drive the driven wheels try to propel the car along a straight line, resisting efforts to deflect it from its path. Driven front wheels apply their tractive effort in the direction in which they are steered and on slippery roads this is a definite advantage.

In a fwd car weight transfer to the rear when accelerating reduces traction, and this wheelspin problem led many designers to postulate that fwd was only suitable with low-powered cars. But the Eldorado and Toronado disprove this contention. Under braking the opposite effect occurs—weight is transferred to the front. The same overloading of front brakes and locking at the rear mentioned for front engine/rear drive cars applies, only more so. Modern disc brakes and proportioning systems at least minimize this disadvantage.

A criticism of fwd cars is that it is necessary to corner with power on, and that if power is taken off the car becomes unstable and oversteers. Fwd cars do corner differently but it's hardly a truism with today's designs that instability when cornering is their basic nature. The reverse is closer to the truth. With almost any car there is some change in cornering behavior when the driver accelerates because a tire which is transmitting power cannot generate as much side force as the same tire when it is rolling freely. So a cornering tire runs at a larger slip angle when also delivering power. With rear-wheel drive, acceleration when cornering increases the slip angles of the rear tires, increasing oversteer (or in the case of most front engine/rear drive cars, reducing understeer). With fwd it is the front tires which run at larger slip angles when cornering under power, so the front end tends to run wide in an understeering attitude with power applied. However, if power is suddenly released the front tires are relieved of their double duty and the car understeers less, assuming the tucked-in nose characteristic of fwd cars under these conditions. Depending upon how much the designers of the particular car have tried to suppress fwd understeer, the result will be simply less understeer or—rarely—some oversteer. But many front engine/rear drive cars also oversteer under the same conditions.

Many manufacturers mention the "arrow principle" in explaining inherently superior straight-line or crosswind stability of fwd. Their reasoning is simple; did you ever try to throw a dart, feathered end first, and hit a target with any degree of accuracy? There is a bit of truth of this "logic," we must admit, as evidenced by the unusually stable behavior most nose heavy fwd cars exhibit in a crosswind. But the reasons are a bit more complex than the simple arrow principle implies.

Rear Engine/Rear Drive

ALTHOUGH REAR-ENGINE cars such as the Julian in the U.S. and the little German Hanomag were produced in the 1920s, it was probably the success of the Auto Union Grand Prix cars in the 1930s that laid the foundation for future mid-engine and rear-engine designs and encouraged Dr Porsche to apply the principle to his VW design. The Auto Union was technically a mid-engine design but the car was so large and the enormous engine positioned so far rearward that a distinction between the two engine locations was hardly ever made. The Porsche 356 which made its debut at the Geneva Show in 1948 broke completely with traditional sports-car design. The conventional front engine sports car of the day had its engine

Quick steering and the ability to hold a controlled power-on drift gained the front engine/rear drive Opel GT a second place in the lane-change maneuver.
Results in the slaloms were as expected: the 914, with low polar moment and good balance, leading; Saab and Opel in the middle of the pack; and the Ghia at the rear.

up front and drive wheels in the back, a ladder-type frame with low torsional rigidity, rock-hard suspension with little roadholding on anything but the smoothest of surfaces, and little or no protection from the elements.

The objective of the Porsche was more along the lines of a high-speed touring GT with due consideration for fast, safe touring in relative comfort. To achieve this goal Porsche used what were basically his VW components—a lightweight, air-cooled engine at the rear, a rigid platform chassis, independent suspension at all four wheels—in an aerodynamic body shape. Compared to conventional sports cars the Porsche had light, responsive steering, smooth ride on all surfaces, a roomy passenger compartment and adequate luggage space, and was incredibly quiet at speed. It also oversteered to an excessive degree because of its rear weight bias, swing-axle rear suspension and skinny tires. And from these early examples grew the notion that all rear-engine cars oversteer in dramatic fashion.

More than 10 years after the introduction of the 356, the first and last modern rear-engine American car appeared: the Corvair. Early examples suffered many of the symptoms of classic oversteer but later models, tamed by better design, were enthusiast's delights. While several companies have built rear-engine cars of modern design today, only the smallest Fiats, France's Alpine-Renault A310, Czechoslovakia's Tatra and various VW models stick with the rear engine placement.

There are several advantages to a rear engine location. As with front drive, there's no driveshaft tunneling through the cabin. With a natural weight bias on the driven wheels, wheelspin should be reduced and traction improved in slippery conditions, again much as with fwd. Weight transfer toward the rear when accelerating increases the load on the rear wheels and further improves traction; weight transfer to the front during braking tends to equalize wheel loads for more even braking. Clean aerodynamics, as important for fuel savings and quiet cruising with a road car as for top speed and stability in a race car, are easier to achieve with a rear-engine layout.

Cockpit overheating, an important consideration with a race car, is less likely to arise when the engine is behind the driver.

Offsetting these advantages are the problems with adapting this configuration to more than a 2-place or 2+2 design. A full sedan design puts an inordinate amount of weight on the rear end when the back seat is occupied and a station-wagon load in the rear compounds the imbalance. Though a sloping front end is fine aerodynamically, it invariably compromises the front luggage compartment, so rear-engine cars can't always take advantage of the lack of an engine up front in this way. Air is the logical cooling medium for rear engines, although water cooling has been successfully used. Thus the problem of cooling the engine in a confined location out of the direct airstream arises. A front radiator is a possibility but this solution is costly, complex and wasteful of luggage space. The obvious solution is a small, alloy engine which can be adequately cooled by ambient air, fitted to a small car. Sounds like a VW or Porsche, doesn't it? The abrupt oversteer characteristic of early rear-engine cars is not necessarily a problem with later designs: it is possible by judicious juggling of suspension, tires, tire pressures, geometry, etc, to considerably tame oversteering tendencies, but the problem of sidewind stability is more difficult.

Mid-Engine/Rear Drive

THOUGH MID-ENGINE designs are typically thought of as products of modern racing technology, Gottlieb Daimler placed the engine behind the driver, offset to the left, on his 1886 motor carriage. This was more for convenience than an attempt at design superiority. However, convenience wasn't the reason for the midship engine layout of the Benz Tropfenwagen, also known as the "teardrop car" because of its uncompromisingly streamlined shape. Raced with moderate success in several versions in the 1920s, the Tropfenwagen was overshadowed by the more glamorous supercharged Mercedes and later Mercedes-Benz cars. The significance of this car wasn't lost on Dr

Slow steering, soft front tires, and body roll (concessions to oversteer tendencies) proved the Ghia's downfall in the lane change.

Heavily loaded outside front wheel worked against the Opel.

Understeering Saab was easiest to control in oil/wet skid pad.

Porsche, who made the midship engine/transaxle configuration work successfully in the Auto Union Grand Prix cars. Later, Porsche designed the mid-engine 550 Spyder, the first racing Porsche not derived from the 356 series, which enjoyed a fantastic competition record and is a forerunner of such successful Porsche racing models as the 904 coupe, the 908 and the turbocharged 917/10.

Cooper pioneered the first truly modern mid-engine Grand Prix cars in the mid-1950s and built one of the earliest sports-racing cars, the Monaco. Following Cooper came a succession of successful mid-engine designs—the indecently quick Lotus 23, Lola 70-Chevrolet, Ford GT, Ferrari 250LM and the much-copied McLaren designs of the late 1960s and early 1970s.

Production mid-engine cars have lagged behind their racing counterparts. One of the first was the Rene Bonnet Djet, later called the Matra Djet, introduced in the early 1960s. The Djet was a somewhat ungainly-looking 2-seater sports car—an inauspicious beginning for mid-engine designs to follow—expensive 2-seat luxury GTs like the De Tomaso Mangusta and its replacement the Pantera, the Ferrari Dino, the Lamborghini Miura (first with a transverse mid-engine) and the thoroughly modern Maserati Bora. These are cars that have ushered in a new era in automotive design, and for those with more down-to-earth budgets there are now the Lotus Europa, Porsche 914—the first mass produced mid-engine car—and the recently introduced Fiat X1/9.

The almost complete takeover of racing by cars with a mid-engine configuration can be explained by the term *polar moment of inertia.* For illustration, consider two bowling balls attached at the ends of a weightless bar. If you lift the bar in the middle and try to turn it you will find that the weight of the balls at each end make movement difficult to initiate and stop. Now, however, if the bar is shortened so that the two balls are placed next to each other the system is much easier to rotate, without a decrease in total weight. The polar moment of inertia has been decreased. When the driver, engine, transmission, fuel tank, etc, are all placed between the wheels, the polar moment of inertia of the car about the center of gravity is low; thus the tires can more easily alter the course of the car. Steering is usually responsive and sensitive too. There are other benefits as well: the majority of weight is again on the driving wheels but the distribution isn't so extreme as with the rear engine or front drive; cornering power can be at a maximum; handling characteristics can be tailored with few compromises. Aerodynamic considerations are an important factor. Unhindered by an unwieldy chunk of cast iron at the front end, designers have far more latitude in shaping the nose and overall body shape for minimum drag with maximum downforce.

But if the driver can easily move the car out of a straight line, so can other kinds of forces, such as a bump or a sidewind. So the mid-engine car gives but it also demands. In exchange for higher cornering power, the mid-engine design asks for a driver with a high degree of competence. Mid-engine cars don't break away easily but when they go, they go suddenly. Drivers who are accustomed to tire squeal or body roll as signs of imminent danger often have difficulty in predicting the breakaway of mid-engine cars because they approach their limits so undramatically.

In racing, where lower lap times alone are the telling factor, a designer may often compromise some areas of the car to reach this goal. Road cars, however, must meet more complex automotive needs. So there are several problems which have kept the mid-engine car from achieving the popularity one might expect. Cost is a primary consideration. Innovation usually costs money, at least until economical solutions to basic problems are achieved.

Because the engine sits right behind the driver in a mid-engine car, such a layout is really only adaptable to 2-seat sports and GT cars. Attempts that have been made to provide extra seats, as in the Lamborghini Urraco, are incomplete solutions at best. For a mass-produced car serviceability is a definite consideration, and accessibility is not a strong point of most mid-engine designs. Add to this heat and noise from an enclosed engine compartment close to the passengers, and you have problems that make even strong engineers cringe. Finally there's the question of rearward vision. No mid-engine car yet with the possible exception of the Porsche 914 has come up with a successful solution to that problem.

The Test Cars

MATCHING FOUR cars of such varying designs for our testing was not the difficult task one might imagine. The Porsche 914 and the Saab Sonett (mid-engine and front-drive categories) were easy choices as they're the only examples available to us at the moment. Our 914 was the 2-liter variety and therefore the most powerful car in our group, but this was of little consequence as none of the tests we planned placed any emphasis on engine performance. Selecting a conventional sports car required a bit more thought, but the choice was finally narrowed down to the Opel GT which fit within the weight, power and dimensional specifications of the previous two choices. Picking a rear-engine sports car was the most difficult. There's the Fiat 850; but it's considerably smaller and lighter. At the other end of the spectrum is the Porsche 911, a heavier car with power and sophistication that clearly set it apart from the basic nature of the rest of our group. Right between these two, however, is the VW Ghia—not a sports car in the strict sense of the term but right as far as size, weight and uncomplicated design were concerned.

The weight distribution of each of these cars pretty well fits the classic definition: the front-engine Opel with 55% up front, the 60/40 distribution of the Sonett, the aft-heavy Ghia with 58% on the rear wheels and the 47/53 distribution of the mid-engine 914.

To eliminate any inherent unfair advantage the 914 might enjoy because of wheel width (the stock 15 x 5½-in. wheels of the 914 2-liter are the widest in our group) the stock cast alloy wheels were replaced with the identical 15 x 4½ steel

Mid-engine 914 has excellent transient response up to the limit but ...

rims used on the Ghia or 1.7-liter 914. And since original-equipment tires vary considerably between the four cars, we standardized the tires. To be sure we were testing configurations and not tires, we obtained a spare set of wheels for each of the cars and fitted each with Pirelli Cinturato CF67 radials of 165-mm section width. Besides the tires, Pirelli supplied one of their top tire engineers, Clive Castell, from their Reno, Nevada test office. His technical advice, recommendations and physical assistance were of great value in our testing. To eliminate tire pressure as a factor in our tests, all tires were inflated to the manufacturer's recommendation for light loads. Pertinent specifications of the four cars are given in Fig. 1.

The Tests

THE NUMBER and variety of tests planned meant several testing sites were needed. We started at the Goldmine ski area in Big Bear, Calif., jumped to the Bondurant School of High-Performance Driving at Ontario Motor Speedway for three days, and finally went to Orange County International Raceway in Irvine for low-traction skidpad evaluations.

Fred Goldsmith, owner of the Goldmine ski area, offered R&T the use of his facilities to test low-speed winter traction capability of the various configurations. Each of the cars was accelerated from rest around a gently curving flat section. At the point where the curve ended and the slope began a marker was placed, beyond which the land sloped gradually upward to an incline of 7-10%. Each car was timed to the marker and the distance driven up the hill past the marker recorded.

Snow condition was hard packed, with a light crust affording maximum traction for this sort of surface. Theory points to the superiority of the rear-engine car in these conditions with front-wheel drive about equal on the flat sections but falling behind on the slopes. Mid-engine and front engine/rear drive cars, particularly, should be at a disadvantage and would be expected to be pulling up the rear of the pack. Over several runs the Saab surprised us by achieving a better time to the marker and running farther up the hill. Several factors explain the Saab's performance. First, the lower portion of hill wasn't unusually steep, so adverse weight transfer off the driving wheels was not an important consideration. Look back, however, to the tire pressure entry in the table of general specifications. Notice that the front pressure for the Saab is 3 lb less than the rear pressure for the Ghia. Our counterparts at *PV4 Magazine* have proved the benefit of reduced tire pressures for traction in the sand, so the Saab has an advantage here. As a quick check we reduced pressures in the front of the Saab and the rear of the Ghia to 15 psi, and now the Ghia motored right past the Saab. We intend to pursue the question of tire pressure vs traction next winter in much more detail.

The 914's superior weight balance worked against it here, and it also suffered from the highest tire pressure on the driving wheels of any of the cars. The Opel's poor showing was expected and can be attributed almost entirely to its 55/45 weight distribution.

Short and Long Slaloms

TRAVELING TO warmer weather, we arrived at Ontario Motor Speedway. Here we devised two tests of transient response: a high-speed lane-change maneuver and low- and high-speed slaloms. Purposes of the slaloms were to determine if transient response of the cars varied with speed and to pick out "overshooting oscillations," if any, leading to instability and oversteer. Pylons were positioned in a straight line 50 feet apart for the low-speed course and every 100 ft for the high-speed slalom. The time from first pylon to last was recorded and later converted to the speed figures listed in the results, tabulated in Fig. 2.

Here the results substantiate theory. The mid-engine 914 with its low polar moment, quick and precise steering and good weight balance proved superior at both speeds. Through the

Fig. 2 TEST	Opel GT f/r	Saab Sonett f/f	Porsche 914 m/r	VW Karmann Ghia r/r
Short Slalom (pylons spaced 50 ft apart, measured distance = 500 ft): speed, mph	27.7	27.8	**28.7**	27.2
Long Slalom (pylons spaced 100 ft apart, measured distance = 700 ft): speed, mph	50.2	51.7	**52.0**	47.5
Lane Change: speed, mph	62.5	61.4	**65.2**	59.5
Snow Traction:				
Time to marker (standing start), sec	20.67	**13.05**	18.78	16.50
Distance driven past marker up incline, ft	10	**143**	24	101
Steady-State Cornering (low traction oil/wet skidpad):				
Speed on 85-ft radius, mph	22.4	**23.4**	23.1	22.7
Lateral acceleration, g	0.396	**0.432**	0.419	0.405

All tests conducted on Pirelli Cinturato CF67 tires, 165-mm section

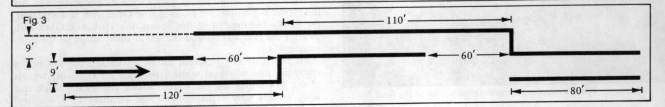

low speed course the excellent caster action of the 914's steering was an important factor in its quickness, and through the long course its balance came into play. It achieved good bite front and rear, with the tail end hanging out slightly.

In a virtual tie for second place through the 50-ft course were the Opel and slightly quicker Sonett. Though the positions remained the same in the high-speed event the margin of separation opened up considerably in favor of the Saab. The Opel's quick rack-and-pinion steering allowed the car to be precisely positioned from pylon to pylon, but working against it was body roll and weight transfer to the outside front wheel. This became very apparent at higher speeds as the noseheavy Opel scrubbed off speed with each reversal of the steering wheel. Perhaps a better-handling front/rear car would have done better.

The Saab's fwd pulled it around the pylons into a second place finish. There is an obvious noseheavy feel to the Saab, but understeer never became excessive. That the Sonett's cornering attitude can be controlled by the throttle played a part in its quickness around the pylons: by proper applications and reduction of throttle the Saab could be aimed from pylon to pylon with precision.

The Ghia ran dead last. Its slow steering, excessive body roll and soft front tires (compromises to reduce oversteer) work against it all the time. The helm does not readily answer the call when asked to reverse direction abruptly, so one must drive slowly to keep from getting caught up without enough steering lock. The heaviness of the tail is evident but never a problem; when it starts to slide it goes very slowly and in a controlled fashion. There is no "dreaded oversteer" with the Ghia, but perhaps if there were it would have done better in this test.

Lane Change

THE LANE-change maneuver was another matter entirely, and though final results are identical to those in the slaloms with one exception, some previously undetected handling characteristics surfaced. This was a test at freeway speeds that stimulated a car's behavior in moving around an object that suddenly blocked its path. Braking was not allowed. Each car was driven between a row of pylons set 9 ft apart. At the end of a 120-ft straightaway marking the entry lane the car had to be jogged left into the adjacent 9-ft lane, then following a short straight section driven back into the original lane (Fig. 3). The time through the course was recorded and later converted to the speeds given.

In the slaloms the quickest times were recorded with smooth driving; early mistakes had a tendency to become magnified later in the course. For the lane-change maneuver proper positioning was less of a concern; avoiding "an accident" was the primary consideration. As a result, as entrance speeds increased, some interesting cornering attitudes appeared. The 914 was again quickest, but not before giving the drivers a few bad moments. The mid-engine car tends to be quick but sneaky: if a driver isn't careful, the limit of adhesion creeps up on him and before he realizes it he finds himself motoring backward. Transient response and balance are excellent up to the limit, but one step beyond and you've got trouble—it takes a skilled and sensitive driver to realize he is approaching the limits of the car, but it should be remembered too that the mid-engine's limits are higher.

In second place, but quite a bit slower than the Porsche, was the Opel GT. Quick steering allowed the Opel to be driven fast down the entry lane and abruptly jogged to the left. A bit of speed was scrubbed off by the heavy front end but by staying on the power we could drift the rear end out in a gentle curve. As the car approached the limit the tires started to squeal and the driver could sense the rear end starting to break traction. By letting up on the throttle slightly we could maintain the drift down the short straight and then reverse for a quick exit.

The Saab could be driven into the first turn with wide-open throttle; once onto the straight section reduced but constant power was required to straighten the car for the quick right.

Full power caused the front end to drift wide, resulting in the driver's clipping pylons on return to his original lane. Releasing the throttle to induce an oversteer attitude like the Opel's didn't work; precious tenths were lost as the car slowed down.

Pulling up the rear was the Ghia—a poor fourth. Slow steering, soft tires and body roll were the Ghia's downfall once again. Steering response is so leisurely there just isn't time for it to catch up with itself after the initial input. The tail meanwhile is swinging out gently, leaving the driver inadequate distance to correct and prepare for returning to the original lane unless he reduces speed. The Ghia oversteers, but in slow motion, so even an unskilled driver can hardly fault its "high-speed" handling.

Low Traction Cornering

THE ADVANTAGE of fwd in limited-traction situations was again proved on the oil/wet skidpad at Orange County International Raceway. An 85-ft-radius circle was laid out and each car driven around as fast as possible. The Saab's steering is without much feel on center but the car maintains an understeering attitude under power at all times, which contributes to ease of control. Other factors also contribute to the Sonett's speed: driven front wheels apply traction in the direction in which they are steered. This and weight on the driven wheels are definite assets in slippery conditions.

The 914 exhibited the same traits here as in the lane-change maneuver—neutrality up to the limit but abrupt oversteer beyond. It places second here because its better front-to-rear balance when cornering imparts an advantage that wasn't usable when driving in the snow.

The Ghia has rear-wheel drive and a decided rear weight bias on its side in this test. But it comes in third for the same reason the 914 does well—balance. Its typical cornering attitude was with the tail hung out; completely catchable, but keeping it caught took valuable time.

There was no quick way to drive the Opel. With the low weight it has on the rear wheels, driving it was a constant battle between understeer with light throttle and oversteer if a bit too much power was applied. The tail didn't come around quickly like the 914's but the constantly changing attitude resulted in slewing and slow lap times.

Side-Wind Evaluation

ONE OTHER test, a side-wind evaluation, turned out to be more subjective than objective because the wind machine R&T rented turned out to be more of a bust than a gust. Luckily a brisk wind gusting to 70 mph at times (no kidding) blew in from the desert for a few days while we had the cars. Our subjective judgments: the front-heavy Sonett with fwd pulling it along was most stable. Next came the also nose-heavy Opel. The 914 was fairly wind sensitive, not a great surprise considering the Porsche's low polar moment of inertia and slight rearward weight bias. Rear-engine cars with light front ends are typically affected more by winds than other configurations, and the Ghia did not disappoint us.

The winner? There is none. This was a test of configurations, not cars; and as our testing proved, each design has specific strong points and limitations. We've laid a few myths to rest: namely, that rear-end cars oversteer drastically and also have better traction in difficult conditions. On the other hand it was satisfying to see many theories proved out by actual testing.

In conclusion, a few conclusions. If you want ultimate handling for driving fast on winding roads—entertainment and pure speed, particularly on dry roads—the mid-engine layout is for you. If you have a need for top traction in the snow where you live, a front-drive or a rear-engine car makes sense and the edge seems to go to the front-drive car. The conventional front-engine car with rear drive seems not to have any compelling advantages, but there are plenty of these cars with more-than-acceptable handling—and you just may have to settle for one anyway, since there are no mid-engine sedans and only one front-drive sports car.

TOMORROW'S CAR

TODAY
OPEL'S CAR FOR THE '80'S

If there is one car money cannot buy, it's this running experimental vehicle, the Opel GT 2, highlight of the Opel stand at the 46th Frankfurt International Car Show.

With this study of the design department, Opel remain faithful to the tradition of displaying technically advanced cars to influence later volume production. Along these lines, the GT 2 joins as the latest in a series of vehicle studies from Russelsheim, as for example the first prototype of the GT with its characteristic rear end, the elegant Diplomat CD and the experimental safety vehicle OSV 40, based on the Kadett.

Says Opel design director Henry G. Haga: "Our target was not only to come up with just a technical feat but to find solutions which would even better adapt the automobile to the human being. Therefore, the primary emphasis of the GT 2 is on high efficiency.

"Efficiency in this sense means that an attempt was made to take fully into consideration functional as well as aesthetic requirements, with attention being paid to optimal active and passive safety."

It thus is a package for the changing life styles of the '80's where energy conservation, crowded driving conditions and the need for more efficient use of space and higher comfort will be clearly in the foreground.

Says Haga: "Due to optimal aerodynamic designing a body shape was found guaranteeing low fuel consumption, best cross wind stability, utmost operation comfort, excellent handling and ideal weight distribution, as well as ample space for occupants and luggage. Consequently, aerodynamic emphasis was not placed on the highest possible top speed but rather on functional efficiency.

"The drag coefficient of the GT 2 is only 0.326 and is therefore 16.6 per cent better than that of the Opel GT, which already had shone brightly for its favourable aerodynamic data."

As in the new Manta, the car's rear end is low, resulting in better aerodynamics and also making for better road hugging. The entire spoiler acts as a functional protective plastic bumper. It incorporates the air intake vents to cool the engine and for air conditioning.

To reduce turbulence in the wheel area, the wheels were designed according to aerodynamic principles. The GT 2 rolls on 6½-in. rims and series 50 (225/50 VR 15) Pirelli tyres.

The roof and the body form a protective cage which, with integrated door beams, provide optimum impact protection. This takes over the function of the roll over bar. In minor accidents, protection is provided by the bumper of plastic foam, which continues through the lower side panels, guarding the vehicle from all sides.

The headlamps are covered and open automatically at the press of a button. The tail light units are matt black to eliminate irritating glare and reflections in daylight. As soon as they are activated, the rear end lights show their normal red, yellow, or orange colours.

Wide sliding doors, opening to the rear, allow easy access to the interior even in narrow parking spares. The sliding doors also provide protection in the event of a side collision. The rear view mirror units and door handles are integrated into the sheet metal of the doors.

The large glass areas are integral with the load carrying body structure. For outside communication, a small drop window within the door window can be opened electrically.

A large glass hatchback allows easy access to the storage area.

The interior, equipped with velour in dark grey and black, is easy on the eye. Pedals and gear shift are optimally located and the steering wheel is height adjustable.

The instruments features electronically controlled digital readouts for speed, engine revolutions and temperature. Further information can be called up by the driver at the push of a button.

The seats consist of bolsters of foam rubber varying in density from soft to hard for optimal body support and comfort. Lateral air spaces between the bolsters improve ventilation.

In addition to the tailor-made driver's seat, the GT 2 has three additional comfortable seats. When the rear seats are not in use they can be individually folded back into the trunk compartment. The interior is fully air conditioned.

All glass areas are tinted and, thus, heat absorbing, with electrical heating for front and rear windows.

The GT 2 is powered by a 1.9-litre fuel injection engine with Bosch L-Jetronic which generates 105 bhp DIN. The powerplant is located behind the front axle, shortening the overall length of the vehicle without infringing on passenger comfort.

The chassis with its tread and camber constancy taken from the new Ascona/Manta, is said to guarantee excellent driveability and high driving safety.

A ZF-five-speed transmission contributes to favourable fuel consumption figures. In this respect — as on the new Ascona and Manta — the result of wind tunnel testing is reflected. Top speed exceeds 200 km/h.

Easy serviceability was also an important design target. All body panels are designed as separate, unconnected units. Not only is this functional but it also allows replacement and repair without the need to disassemble and reassemble adjacent parts. All instruments are simply plugged into sockets for easy service.

Opel do not intend to put the GT 2 into production. It represents a car whose progressive ideas might be found in future series production cars of Opel. It is an experimental vehicle in which new developments can be tested and perfected, developments which will better adapt cars in the future to man and traffic●

The Opel GT

Jerry Sloniger chronicles the silver dream car which became reality and changed Opel's image.

OPEL has applied the "GT" title to a good many models in its postwar history, particularly during the past 15 years, in a successful attempt to shed their grandpa image in favour of a sporty-youth flair. But only one model from Opel was a true GT in the classic sense of such initials, and that came about almost despite the intent of the firm's biggest bosses.

An American Chief Designer conceived the Opel GT precisely to turn their belt and braces image around, and he found a willing confederate in the German head of engineering. Before anybody could figure out why not, Opel had its GT in form as well as name, a descendant of the show car syndrome which escaped the clutches of the spotlighted turntable and sequined forms to grace an open highway.

Our tale begins with a silver dream car at the biennial Frankfurt motor show in 1965, the one occasion when *every* German car firm must grab local attention. But this was only the tip of the first iceberg.

Clare MacKichan of Styling, or Design as it came to be called, and a gentleman named Mersheimer, decided that Opel's need was a two-seater, a sort of "continental Corvette". Starting with a Kadett chassis, somewhat modified, they built just that. But it differed from other Opel dream cars in two important ways:

First, this Opel GT, introduced at the end of 1965, was a proper runner, even a hot one. Secondly, it led directly to "production", although that would take another three years. To anticipate our tale somewhat, we must admit it also remained unique, although at least two more attempts were floated to keep Opel in the GT game.

Their first show car, dubbed "experimental GT" in a brochure then, had a tuned engine with four carb throats jutting purposely forth, and it had what Opel was forced to call a "revolutionary" back axle arrangement — meaning coil springs and four locating links. In fact, this was their chosen new Kadett (B-range) axle for 1967 so any spin-off into a GT would have to wait at least that long to appear on the open road.

In the meantime Styling sold Engineering the idea of a GT with its engine moved back nearly seven inches in the Kadett frame. The original goal was a low, sloping bonnet but the argument which won was "better handling". This was proven by building one car each way and having their best demon driver of the middle-'sixties thrash them around Nürburgring without knowing which was which.

He chose the rear-set engine by a country mile, and proved it by lapping the Ring much faster. As a bonus, the configuration also moved their shift lever back to a near-ideal position for snap shifts, although their GT would also be one of the first Opels with the option of their new Strasbourg-built automatic.

The original or show GT had a very thin air intake and longer tail than the eventual road car. In the interim, Opel had put it into a wind tunnel, but also added 2·75in of headroom with a larger windscreen and wider interior. The aerodynamicists then dictated the blunter tail as well and they ended with a C_d factor of 0·39, considered remarkable for the middle 'sixties.

Admittedly, there was still virtually no air flow through the cockpit and about as little vision to the rear quarters — but it sure looked mean, which was the point, after all. At least it did to Americans, who bought up to 80% of the original GTs through their Buick dealers, while Germans generally left them alone.

In original form the car came with either a 55bhp Kadett four of 1·1-litres, which hardly gave performance to match the price and shape, or an untuned 1·9-litre four from their Rekord sedan with 90bhp (dual carbs and the like had been forgotten). The car weighed 2150lbs by now and took the best part of 11 seconds for 0-60mph runs with normal gearing but the fine shape did allow 115mph top speed and that would keep it ahead of a Capri or Fiat 124 Sport.

Part of the weight came from a very sturdy steel body. Repeated rumours of a glassfibre Opel GT were never more than talk, probably inspired by the Corvette back home. Opel never seriously considered the material in this case.

Even in steel they were having a problem getting the GT built. The main plant in Rüsselsheim had no spare capacity and a projected yearly output of perhaps 20,000 didn't prompt management to find some. In the end, a French coachworks firm name Brissoneau & Lotz outside Paris, built the GT.

The sporting people at Opel soon realised that the car needed more power, or at least a little panache, early in its career. When I sought a proper weapon for a pair of short rallies in May of 1969 Opel practically leaped at the chance to build a rather special GT (in both Vienna and the hills behind Monaco, any tune was fair). They couldn't do it directly of course, but Irmscher could.

At that time the Irmscher empire — today the sole recognised Opel tuner — was hardly begun, so deniability was built into our project. Irmscher cut away some wheel arch and stuffed in a pair of dual-body Webers to go with 11·2:1 compression (normally 9·5). A total of three cam bearings limited changes there (and revs to 6000) but the bottom end was bulletproof so we had pistons lightened by 200g. In all, the now-Group 6 machine produced 150bhp — and a blow-back fire.

Above, Gallion's model of the rarest of Opel GT's, of which he owns the second full-size example. Below, prototype with chisel nose.

Below, the same prototype on its stand at the 1965 Frankfurt Motor Show.

The Opel GT

With no more chassis tuning than Bilstein shock absorbers and a 4·22 final drive the car still cleaned up on also-new Capris and even took its class from Porsches at both events, proving it was fast and stable — whenever you could get the power down.

Henri Greder, then deep in Opels generally, was already working with Buick to take out 450lb, lower the GT by 2in and fit 7in rims but this project trailed off. Then Opel dealers in Italy sold Alfa-wizard Conrero on the idea of a GT all their own. He added 2mm of bore, to 95 and a full two litres, balanced it all carefully for 8500rpm and got as much as 190bhp for short events, with a 150mph potential top speed and 6 second 0-60 times. One finished 9th overall and first in the GT class at the Targa.

But it still wasn't catching enough German attention. Perhaps it wasn't promoted with full enthusiasm since many at the top of Opel's tree couldn't entirely accept a "French-built" machine of limited interest. Styling did something about the latter in 1971 by introducing the GT/J with matt black replacing chrome for trim and the 1·9 engine standard. Eventually they even sold a fifth of all GTs in Germany and now have a club for the car there.

It was always popular for special duties, particularly record-setting in curious classes. In May 1971 the dealer Georg von Opel (descended from the original family but not active in the GM-owned firm) assembled a pair of Bosch electric motors and 1260lbs worth of a new Varta nicad aviation battery, stuffing the lot into a GT. Pumping out the tyres, he took 6 world electric records ranging from the ¼ mile drag at 53·4mph to 10 miles with standing start at 79mph. He had clear plans to fit 1630lbs of batteries and try for 100km and 1 hour records but dropped those.

A year later Opel took over the basic idea but went much deeper. They put a longer nose and one-man bubble top on a GT base, fitted their imminent 2·1-litre Rekord diesel and mounted a precision attack on compression ignition class marks, capturing 18: all the way to 10,000km and 24 hours with 95 turbocharged diesel bhp. The fastest time was 123mph for the flying kilometre.

Meanwhile, for the 1969 Frankfurt show in fact, MacKichan's successor at Opel Design had produced a targa-topped GT called the Aero. Under Chuck Jordan, Opel tried to avoid the t-bar top by bracing the lower body and removing the top panel entire, while adding flying buttresses to the C-pillars.

Actual construction was arranged in Italy by Dusio's son (of Cisitalia fame) but it was a poor job with quick plastic side windows and a rear pane which remained fixed under the bar. Its shape had been done so fast at Opel they worked straight from drawings rather than clay in the prescribed GM manner. Opel also introduced a V-8 special called the CD at this same show which was all dream car. The Aero was largely overlooked by press and public.

But not by the organisers of the Los Angeles car show who asked for an Aero. Opel decided to do them a proper one and gave the second car to Michelotti, incidentally as the first of many such projects he would do for Opel. He took over the first car's tooling but devised a set of quarter-circle cog tracks which fold the rear window down under the rear deck. And he got the glass side windows in.

The idea of a windscreen extended into the roof had been abandoned before the first Aero, but they did use frameless side glass of spherical section which must be wound up or down very carefully indeed. The removeable top is held by two pegs in back and two catches in front although it — like the entirely open top — would hardly have survived the rigours of production testing without more braces.

Styling obviously hoped this would be an addition to the Opel sales programme but management had seen all the GTs it cared for. Even so, both Aeros were actual, running machines. The better (second) car did about 11,000 miles in various works hands before George Gallion, then and now second in command at Opel Design, bought it in 1973. He had joined the German studios from Chevrolet in 1969, as the Aero was nearing completion.

Since acquiring this Aero he has used it for another 11,000 miles, painted it once and added the matt black tail panel which many GT owners favoured — but the car is currently as-built, mechanically.

Today he admits Opel would have needed a strut between windscreen and targa bar for rigidity. After a decade you can't fit his top panel if the doors are open and people in the seats. Otherwise it is amazingly solid and just about as muscular to drive as the rally car we ran about the time this Aero was born.

But this, and the less-refined Aero, remain the only two topless GTs in the Opel line, among 103,463 built in five years. Of course a new Kadett was due by 1973 and that would have meant new chassis engineering for the GT as well. Brissoneau had been absorbed by Renault in the meantime too, and the French needed its production capacity. So the GT ideal was allowed to fade away.

Opel Styling wasn't quite through, however. Their GT2 appeared at the 1975 Frankfurt show with a claimed C_d value of 0·326 and an estimated top speed of 120mph on 105bhp. But sliding doors and digital dash indicated even Opel didn't truly mean to build any more of those.

No matter how sleek the Aero might have been, Opel was getting ample youth image from its regular models and "dealer" rally teams. A limited-production, overly-specialised car like the GT was no longer needed. So they simply didn't build any more. Especially in Aero form, the Opel GT was admired and happily driven on special occasions by a great many Opel leaders — but it never quite seduced enough of them at any one time to produce an heir. ●

Left, the rare Aero with top in place. Below, Aero configuration far more evident with lid off and seen from above.

Above, prototype engine bay showing "hot" engine. Below, well laid-out cockpit with nice large dials and well placed gearchange lever.

CLASSIC CURVES

Roger Legg tells the full story of the rebirth of his beautiful and rare Opel GT

EARLY in 1982 I sold my trustworthy seven-year old Triumph-based special and set about finding replacement transport. I wanted something fairly 'Classic' that fell within my budget. Many of the sports cars that I fancied were way out of range.

Undeterred I bought that month's issue of *Classic Cars* and as usual scanned the classified ads first. And there it was: *"Opel GT, two-door sports coupé. Needs total restoration, two engines, Cambridge."* It rang a distant bell and reference to a back issue of the magazine (November 1980) came to the rescue. An article called simply *The Opel GT* jogged my memory and filled in some of the background to these beautiful little coupés.

Seduced by those double curves, arrangements were made with the owner, Bernard Montgomery, to view the car next day. The car had never been registered in England and was still on its German import plates, 252-Z-785, although while in Bernard's ownership he had paid the import duty. All he knew of its history was that sometime in 1971-72 it had been brought into England by a serviceman serving in Germany, and once its licence had expired, the car had been laid up. The condition was much as I expected after my conversation with Bernard the previous evening. A price was agreed and arrangements made to get it transported to my home, along with boxes of bits, including spare pistons, con-rods, rings and shell bearings. I should mention that the gearbox was out of the car and the engine was hanging on its mountings.

At this time my intention was to convert the Opel to right-hand drive and enquiries were made as to feasibility, without much success due to everyone thinking I was talking about an Opel Manta. I became more aware of how rare my new car was, and how little people knew about it in Britain.

Three weeks later the car arrived, complete with an Autobooks Manual and a copy of *Style Auto* which covered the production of the GT. So amid remarks like: "You aren't going to drive that — it's lhd and what are you going to do with that?" a closer inspection of what I had bought took place, while pictures of the finished car flashed through my mind.

Our regular transport was relegated to the drive and two weeks later, among boxes of labelled bits, I had a stripped bodyshell sitting on a newly made trolley for manoeuvrability. During these two weeks more information in the form of Brooklands Books' *Opel GT*, Auto Classics' *Opel GT Coupé* and *Nur Fliegen Schöner* had been bought.

Above, part way there, but still a lot of work to do; Roger Legg's Opel GT after partial shotblasting and with remetalling in progress

I then discovered just how rare my particular car was, being a GT-A-L Deluxe Model of '79, of which, out of a total production of 103,463 only 3,697 had been built. This called for a change of plans and the rhd conversion was forgotten, so along with photos and copious notes made while dismantling, a list of parts required for the rebuild was drawn up.

I then called in at the local mobile sandblaster and together we went round the shell blasting what we thought were the rough areas; bottom of both doors, ends of the sills where they met the wings, the flared arches and any other dubious looking areas. There weren't many surprises and the underneath was surprisingly sound, due probably to the car having spent most of its life standing under cover.

The following Saturday, armed with my parts list, I arrived at my local Opel dealer, and there I met Phillip Cross whom I now consider to be a good friend, although at the time he took the smile off my face with those immortal words: "Oh God, not one of those". At his suggestion I left the list. Looking back, what a list it was! I arranged to see him again the following Saturday. Saturday duly arrived and the smiles returned, for Phil had a list of part numbers and prices. To our surprise the majority of the parts were in stock at Luton. The rest it appeared would take about three weeks to come from Germany. So, parts on order, it was back to the garage and armed with sheet metal and welding torch, the rebuild began.

Although the shell wasn't in bad condition, time was consumed remaking the damaged flared arches and making part panels to fit those double curves. What I would have given for access to a wheeling machine! Also time consuming was putting right two areas of accident damage which had been filled. The offside door had several creases along its length and took a lot of careful planishing to bring it back to original. When all was welded and dressed, it was time to lead load the repairs and, never having done this before, I had more on the floor than on the shell. I needed help! Contact was made with an old Rover employee, Sid Holden, who came over and spent the day loading my repairs and teaching me the art; was I grateful for the help of an expert! Finally the repairs were finished, the rest of the old paint removed and the shell prepared and repainted inside and out in the original Monza Blue.

By this time eleven months had elapsed, so with the body put to one side and duly covered it was time to start on the mechanicals.

The engine, which according to the indicated mileage had done only 72,265km (45,165 miles), was stripped first. To my dismay, when the head came off, the bores were empty and badly

CLASSIC CURVES

rusted, as was the crankshaft, discovered minutes later. The so-called 'spare' pistons and con-rods that came in one of the boxes were in fact out of the engine. Having suitably cleaned all the bits I hauled the block, head, crankshaft and pistons off to another new friend, Mike Stevenson, for his expert advice. It was decided that a rebore, crank grind and head reface were required and that Mike would rebuild the crank and pistons back into the block while I prepared the ancillaries.

The rest of my time was spent stripping the rear axle, front suspension, steering and making another list of further parts required. Incidentally I would advise anyone who is removing the front transverse leaf spring to make a spring compressor, if you can't borrow one. I found to my cost when it let go; you can't move fast enough! Besides which you will not be able to refit it without a compressor.

While awaiting the parts, all the hardware was taken to Redditch Shotblasting who did a great job of removing all the old paint and leaving a superb finish ready for repainting.

All the hardware was primed with Corroless Primer, which has a rust inhibitor, and then sprayed with Gipgloss Nu-Coat black, which can also be brushed and that can be an advantage if any touching in is necessary after rebuilding. No major problems were encountered with the rebuild of the running gear except for the front upper wishbones, where it was found that no bushes were available. In the end I had these specially turned out of nylon. Bushes are now available through Jan Lisewski at Parsget UK, Derby, who was one of the founders of the Opel GT UK Owners Club.

Twelve month wait

The running gear finished, it was time to uncover the shell which was brought out of its cocoon of twelve months. It was liberally retreated with Waxoyl, left for a week to drip, and the completed running gear fitted over the next few weeks. The only diversion from originality was the fitting of copper brake pipes, a necessity in the UK climate.

Next came the wiring which, although fiddly and time consuming, thankfully had been fully labelled when removed. This was cleaned and the continuity and connections checked with battery and bulb before being refitted into the shell.

While the engine compartment was still empty I decided to fit the revolving headlight pods and mechanical turning gear. After what seemed a lifetime, and a lot of cursing, I was satisfied to find that the pods turned and locked together; what a relief. The bonnet was also fitted at this stage as one could sit underneath and adjust the bolts and locking catches easily.

Mike had finished the engine sometime before and this had been rebuilt with most of its accessories ready for installation along with the gearbox. This was a tricky operation as the front of the car had to be raised three feet and the engine slid underneath on its body mounting frame.

With patience and steady juggling, it was bolted in place along with gearbox and propshaft. Once back on *terra firma*, all electrical and hose connections were made; so, after 25 months, the car was taking shape.

The temptation to rush the finishing stages was overcome and attention turned to the facia and interior. Although most of the interior was in good condition and only replacement of the inner sill, rear deck carpet, vinyl covered back panel and parcel shelf were required, my next problem arose in trying to match the colour, blue in this case. Eventually Marsdens of Birmingham came up with a vinyl cloth and carpet of almost exactly the right shade. After several evenings behind a sewing machine, the carpets were finished and a new gear lever gaiter made. The rear panels were covered in the vinyl cloth and after fitting the shade difference isn't noticeable.

Beyond saving?

The facia, which was badly cracked, was taken round several coachtrimmers who looked at it in despair and said the only satisfactory repair would be to stitch a new cover over it, but this would have left visible stitching lines which I didn't want. I decided to fill the cracks with plastic filler and repaint the whole facia with Humbrol Vinyl paint which one coach trimmer recommended. I had to mix the paint to the correct shade; but when finished the facia looked terrific and so was very carefully fitted into the car. Alas three days later the cracks were reappearing. I decided to give up the attempted repair and seek a replacement. A 'phone call was made to Jan at Parsget and several weeks later a secondhand facia had been found which was perfect, although black in colour. After checking that it would fit, it was sprayed blue, and two days later installed in the car; it looked like new. This was followed by careful fitting of the instrument panel and the upper steering column.

The interior was by this time complete except for the seats. The doors were refitted at this stage and, although the hinges had been marked for positioning, they took several hours of juggling to arrive at the correct gap all round the opening.

The front and rear screens were fitted with new rubber seals and installed into their respective apertures, followed by the opening rear quarter lights. The door glasses were fitted to their lifting mechanisms but when installed would not lift square to the frame as they had done originally. The reason for this was traced to the nylon guides on the lifting track which had been cleaned and Vaselined for ease of operation. The weight of the glass was showing up the wear on the guides. The lifting cranks were removed and after some juggling with the nylon guides, were refitted into the doors whereupon the glass wound up and down in parallel. The door latching mechanism was fitted and attached to the locks, followed by the door trims and the seals.

All the lights were then connected and fitted into their openings; those at the front being converted to left hand dip. With the battery fitted, a test of all the electrics was carried out. Much to my surprise the offside dip beam wouldn't work; this was eventually traced to a length of perished wire running three feet from the bulb holder into the loom. Although working when tested, it had parted company when fitted back into the car. After splicing in a new wire, all the electrics were working and the time for a major test had arrived. Would the car start and move under its own power for the first time in 15 years?

The spark plugs were removed, oil pressure built up and the carburettor primed; plugs back in, and several coughs later it fired and ran, but very unevenly. Although at first the carburettor was suspected, a check through everything revealed such a simple mistake. Two of the plug leads had been opposed in the distributor cap. Once fitted correctly, the tickover was adjusted, first gear selected and my Opel moved out of the garage under its own power.

Finally I fitted the grilles, bumpers, wheel trims and badges and bravely made an MoT appointment for the following week.

Having never driven an Opel GT or a lhd car before, the first couple of miles were taken very steadily. Thankfully it was Sunday and there was little traffic. I immediately felt at home in the driving seat. Happily the Opel sailed through the MoT. Certificate in hand, the journey home seemed over much too fast.

Thanks to the Local Licensing Office, and confirmation from Opel in Germany that the car had been built on September 8, 1970, the car was registered with an appropriate registration for its year of manufacture.

The cost of the rebuild was approximately £2,300 not including my own hours of labour. I would like to thank all those people not mentioned who gave advice or help during the restoration and most especially my long suffering wife, Janet, who has no interest in cars except for getting from A to B.

My first major outing in the Opel was to Shelsley Walsh Hillclimb, about an 80 mile round trip, in preparation for a journey to Brands Hatch and entry in the National Classic Car Concours. This event was followed by the UK Owners Club GT Spectacular at Silverstone, where a large contingent from European Clubs had come to join us for three days. It was a great success and I was delighted by comments from several of our continental friends who were amazed at the originality of the rebuild as they tend to add on all the original extras like rear window visors, wheel arch extensions, air dams, spoilers etc.

For anybody contemplating a similar project the following tips may be of use:—

1 Don't fix a deadline because it usually takes longer.

2 When stripping the car make notes and take photos of everything so you have a record of where and how it was fitted. This is a must if information and your make of car is scarce.

3 Label everything as it is removed.

4 Look at what you are about to do and see if any problems may arise; it's no good if it falls apart and you don't know how it goes together.

5 Work in a logical way especially when rebuilding so you only have to fit parts once.

6 Only do it when you feel like it, so it doesn't become a chore.

7 Join the club, if there is one.

Finally was it worth it? Most definitely 'Yes'; for as the ads of the day read, *Opel GT, only flying is more beautiful,* and I've only done 800 miles so far!

PS. If anyone knows more details of this car's history I would appreciate hearing from them. A clue which might jog someone's memory: while stripping the interior, I found an embarkation card with the name: Edith Brenner, Birthdate 22-10-54, Bürstadt.

The Opel GT UK Owners Club can be contacted through Graham Whitehead, 301 Vale Road, Ash Vale, Nr Aldershot, Hants.

Above, care and attention to detail are the secrets of a truly successful restoration. Stripping and refitting the engine bay of Roger Legg's car took many hours before the rebuilt engine was lowered into place